Minitab® Lab Manual for Vining's
STATISTICAL METHODS
FOR ENGINEERS

Roger E. Davis
Pennsylvania College of Technology

DUXBURY PRESS

An Imprint of Brooks/Cole Publishing Company
I(T)P® An International Thomson Publishing Company

Pacific Grove • Albany • Belmont • Bonn • Boston • Cincinnati • Detroit
Johannesburg • London • Madrid • Melbourne • Mexico City • New York
Paris • Singapore • Tokyo • Toronto • Washington

Sponsoring Editor: *Cynthia Mazow*
Editorial Assistants: *Rita Jaramillo, Kimberly Raburn*
Production: *Dorothy Bell*

Cover Design: *Roy Neuhaus/Bill Reuter*
Cover Photo: *Mark Romine/SuperStock, Inc.*
Printing and Binding: *Webcom Limited*

For more information, contact Duxbury Press at Brooks/Cole Publishing Company:

BROOKS/COLE PUBLISHING COMPANY
511 Forest Lodge Road
Pacific Grove, CA 93950
USA

International Thomson Publishing Europe
Berkshire House 168-173
High Holborn
London WC1V 7AA
England

Thomas Nelson Australia
102 Dodds Street
South Melbourne, 3205
Victoria, Australia

Nelson Canada
1120 Birchmount Road
Scarborough, Ontario
Canada M1K 5G4

International Thomson Editores
Seneca 53
Col. Polanco
11560 México, D. F., México

International Thomson Publishing GmbH
Königswinterer Strasse 418
53227 Bonn
Germany

International Thomson Publishing Asia
60 Albert Street
#15-01 Albert Complex
Singapore 189969

International Thomson Publishing Japan
Hirakawacho Kyowa Building, 3F
2-2-1 Hirakawacho
Chiyoda-ku, Tokyo 102
Japan

Printed in Canada
10 9 8 7 6 5 4 3 2 1
ISBN 0-534-23709-6

Table of Contents

Preface

A large volume of research indicates that students remember only a small portion of what they see (as in reading this manual) or what they hear (as in listening to lectures), but retain most of what they actually do. Therefore, the objective of this manual is to have you spend a majority of your time doing the things that you are learning.

Statistics is about asking questions and attempting to provide tenative answers to those questions in a coherent manner. Statistical methods represent a collection of tools which an investigator uses to apply basic scientific principles to real problems in an efficient and effective manner, gaining insight into possible answers to the questions or hypotheses under consideration. The information obtained by using these statistical tools are carefully organized to illustrate patterns which shed light on the hypotheses.

Upon completion of all the material in this lab manual you should be able to:

- Use Minitab to analyze data and produce appropriate graphics for each exercise.

- Ask interesting and reasonable questions about data.

- Select appropriate methods to gain insight into possible answers to those questions.

- Organize and summarize the statistical output.

- Discuss the relevance of your results in relationship to your initial questions.

This Minitab Manual was written to accompany G. Geoffrey Vining's STATISTICAL METHODS FOR ENGINEERS. The examples used in this manual are taken directly fom the text; the examples in this manual appear with the same title as the text. As an illustration, the Minitab Manual contains **Example 2.5 Sensory Modalities** in the section entitled Boxplots (Parallel Boxplots). This example corresponds to **Example 2.11 Sensory Modalities** in Vining's text in the section entitled Boxplots (Parallel Boxplots). This Minitab Manual contains specific step-by-step instructions enabling the reader to produce the statistical output corresponding to the examples in Vining's text. This Minitab Manual might best be used in conjunction with Vining's text in a laboratory setting or in supplementary assignments for the student. Minitab 11 was used in the preparation of this manual.

The data files for STATISTICAL METHODS FOR ENGINEERS by G. Geoffrey Vining are available with the text book and on the website: www.Duxbury.com. The data files are named according to the chapter and exercise number to which they correspond in the textbook. For example EX0731.MTP is the Minitab portable data file for exercise 7.31 in the textbook. References to those data files, in the same format, are used in the exercises in this Minitab Manual.

Words that the student should type in are printed in italics. For example, the instruction: Name column C1 by clicking on the cell above the first row and typing *Before.* requires that the student type *Before.*

Chapter 1
Introduction to Minitab

1.1   Overview

This chapter covers the basic structure and commands of Minitab for Windows Release 11. After reading this chapter you should be able to

- Start Minitab.
- Identify the Data and Session Windows.
- Enter Data into Minitab.
- Save the Data File.
- Compute Descriptive Statistics.
- Print the Session Window.
- Obtain Online Help.
- Exit Minitab.

1.2   Starting Minitab

Minitab is available on the network in the various computer labs. The procedure for starting Minitab requires that you:

1. **Select Start>Programs. Locate and point to the Minitab program group** as shown in Figure 1.1.

2. **"Double click" on the blue Minitab for Windows icon,** as shown in Figure 1.1.

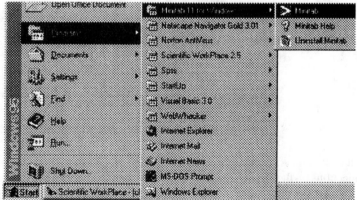

Figure 1.1

The main Minitab window, as shown in Figure 1.2, contains four subwindows: the Data window, the Session window, the Info window, and the History window.

Across the top of the Minitab window is the menu bar, from which menus may be opened and from which you choose commands. The Session and Data window are the most importand and most frequently used windows. The Info and History windows are normally minimized and not used.

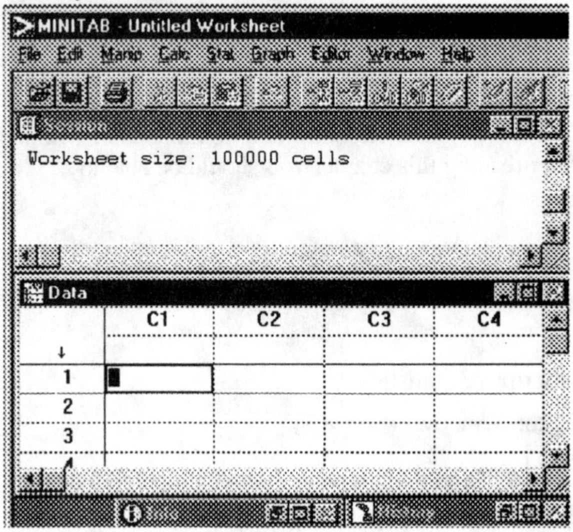

Figure 1.2

The main menu bar, as shown in Figure 1.3, contains selections common to most Windows applications and some selections specific to Minitab. The File command contains options related to opening files, saving files, printing and exiting Minitab.

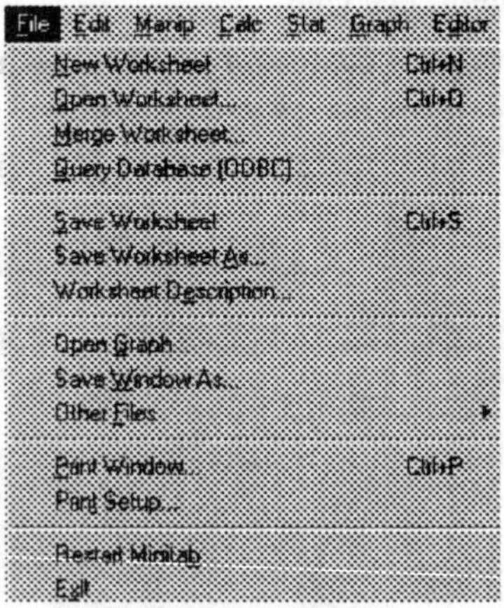

Figure 1.3

The Edit command contains options related to deleting, copying, and pasting. The other selections on the menu bar are sepcific to Minitab: Manip(ulate), Calc(ulate), Stat(istics), Graph. The final two selections on the main menu bar, Window and Help are found in most windows applications. The Window command enables you to switch among windows, while the Help command enables you to get on-line help from Minitab.

Minitab's Data window, shown in Figure 1.4, is like a spreadsheet in that it works with data in rows and columns. Typically a column contains the data for one variable, with each individual observation in a row. Columns are designated as C1, C2, C3... and rows are numbered 1, 2, 3,...

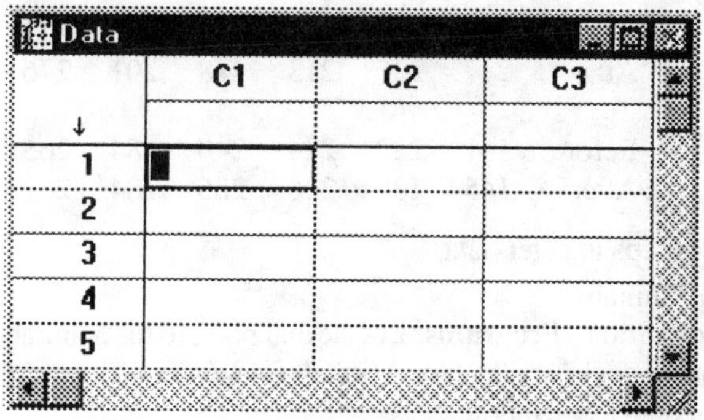

Figure 1.4

Minitab is a computer software program initially designed as a system to help in the teaching of statistics, and over the years has evolved into an excellent system for data analysis.

The size of the worksheet is limited only by the memory available and the size of the hard drive.

Adding Comments in the Session Window

It is often desirable to add comments at appropriate times within the Minitab session. These comments provide a reference when the Session window is printed or edited.

Follow these steps to enter a comment in the Session window.

Step 1. Go to the Session window.

Choose **Window**>**Session** or use {Ctrl} M to go to the Session window.

Step 2. Enable editing.

Choose **Editor**>**Make Output Editable**. When the output is Editable, you can add comments anywhere in your Session window.

Step 3. Enter the comment.

These comments may pertain directly to the Minitab output or other information as directed by your instructor, such as Your Name, Course, Section, Today's Date.

1.3   A Minitab Session

Entering Data into Minitab

There are several ways to enter data into the Minitab Data window. Let's look at a problem involving entering data into Minitab by typing the data into the Minitab worksheet (Data window).

3

Example 1.1

A study was done to determine whether a new process will reduce operating temperatures on aluminum cylinder heads used in a specialized engine. Twelve different engines provided the following measurements.

Before	271	237	212	143	285	273
After	217	235	213	118	201	276

Before	161	222	224	240	184	265
After	145	180	236	286	184	237

Follow these steps to enter data.

Step 1. Start Minitab.

Choose **Start**>**Programs**. Locate and point to the Minitab program group. "Double click" on the blue Minitab for Windows icon.

Step 2. Open the Data window.

If it is not already visible, open the Data window to view the worksheet by choosing **Window** > **Data** or pressing {CTRL}D.

Step 3. Maximize the Data window.

To make data entry as convenient as possible, maximize the Data window by clicking on the Maximize button in the upper-right corner of the window.

Pressing {ENTER} moves the pointer in the direction of the data entry arrow shown in the upper left hand corner of the Data window. You can change that direction by clicking it with the mouse.

Step 4. Enter Data.

Enter the data in the data window. A portion of the Data window is shown in Figure 1.5. Observe that the pointer in the upper left hand corner of the Data window is pointing down, in the direction of the data entry.

Step 5. Correcting Errors.

If you enter an incorrect value, highlight the cell, retype the data entry and press {ENTER}. (Do not delete the error, just type in the correct value. Deleting the data causes the entire column to move up one line!)

Step 6. Name the columns.

Name column C1 by clicking on the cell above the first row and typing *Before*. Name column C2 by clicking on the cell above the second row and typing *After*.

	C1	C2
↓	Before	After
1	271	217
2	237	235
3	212	213

Figure 1.5

Saving a File
Follow this step to save the file.
Choose **File**>**Save Worksheet As**. Select drive A: from the Save in: drop down list box. Type *ex1_1* in the File name: text box. Choose **Save**.

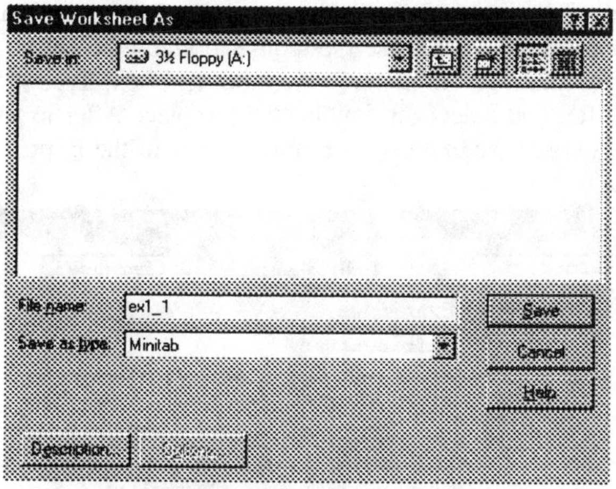

Figure 1.6

Entering Patterned Data
   Let's enter a column of patterned data (1,2,...12) which will be called Code.

Follow this step to enter patterned data:
Choose **Calc**> **Make Patterned Data**>**Simple Set of Numbers**. Type *Code* in the Store result in column: text box, as shown in Figure 1.7. Place 1 in the From first value: text box. Place 12 in the To last value: text box. Place 1 in the In steps of: text box. Choose **OK**.

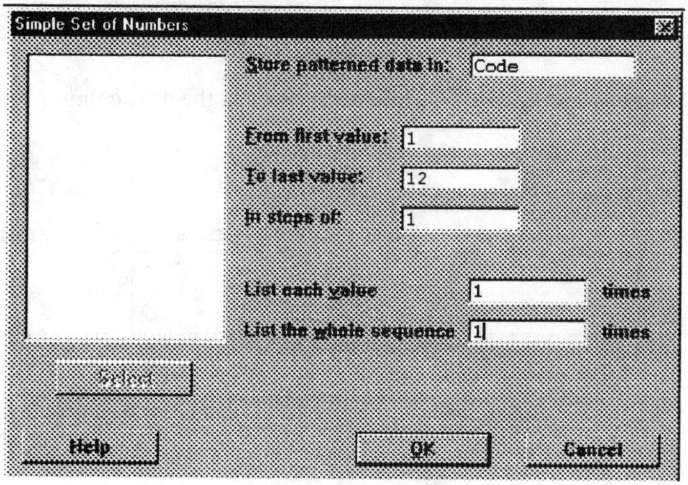

Figure 1.7

Arithmetic Operations on Columns

Determining how much the operating temperatures on aluminum cylinder heads has changed for each engine involves a simple arithmetic operation.

Follow this step to calculate the difference:

Choose **C**alc>**Ca**lculator. Type *Difference* in the **S**tore result in variable: text box, as shown in Figure 1.8. Use the mouse to highlight Before and Select (or double click) to place Before in the **E**xpression: text box. Type -. Use the mouse to highlight After and Select (or double click) to place After in **E**xpression: text box. You should see the expression Before - After in the **E**xpression: text box. Choose **O**K.

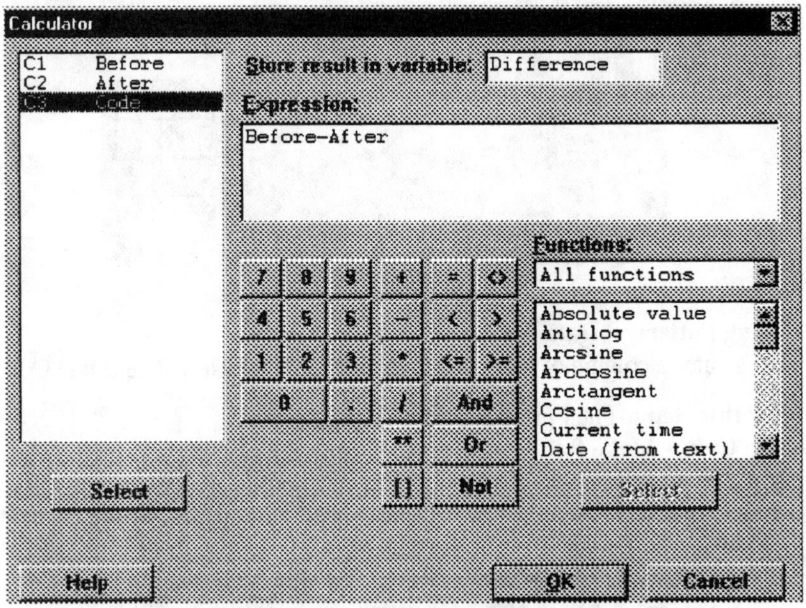

Figure 1.8

Observe that the new variable Difference appears in the next column, as shown in Figure 1.9.

	C1	C2	C3	C4
↓	Before	After	Code	Difference
1	271	217	1	54
2	237	235	2	2
3	212	213	3	−1

Figure 1.9

The Session Window

The Session window displays numeric and text output. Minitab offers a variety of basic statistics to analyze the data. Let's begin by obtaining a summary table describing the three variables Before, After, and Difference.

Follow this step to obtaining the summary table.

Compute descriptive statistics.

Choose **Stat**>**Basic Statistics**>**Descriptive Statistics**. Place the three variables (Before, After and Difference) by highlighting and "double clicking" on
the variables or choosing Select in the Variables: text box, as shown in Figure
1.10. Choose **OK**.

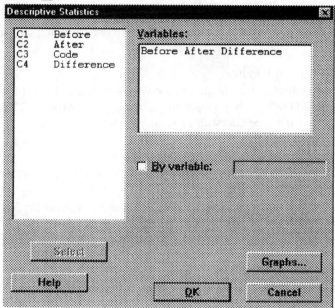

Figure 1.10

Since Minitab is interactive, separate files do not exist for input and output.
The results of the interaction between Minitab and the user are displayed in the
Session window, as shown in Figure 1.11.

Descriptive Statistics

Variable	N	Mean	Median	Tr Mean	StDev	SE Mean
Before	12	234.7	230.5	230.9	59.9	17.3
After	12	210.7	215.0	212.4	49.1	14.2
Differen	12	24.1	9.0	20.7	47.0	13.6

Variable	Min	Max	Q1	Q3
Before	143.0	365.0	191.0	272.5
After	118.0	286.0	181.0	236.8
Differen	-46.0	128.0	-2.5	51.0

Figure 1.11

The values in the section of the session window labeled Descriptive Statistics indicate the number of observations (N), the sample mean (Mean), the sample median
(Median), 5% trimmed mean (TrMean), sample standard deviation (StDev), and
the standard error of the mean (SEMean). The minimum value (Min) is listed,
along with the maximum value (Max), quartile one (Q1), and quartile two (Q2).

Printing the Session Window

Choose **Window** > **Session** or use {CTRL}M to go to the session window.
Follow these steps to print a copy of the Session Window:

 Step 1. Select the printer.

    Choose **File** > **Print Setup** to select the correct printer. After selecting the correct printer. Choose **OK**.

 Step 2. Print the Window.

    Choose **File** > **Print Window** to print the session Window. Choose **OK**.

Saving the Session Window

    A record of the Minitab Session window may be saved in a text file. This text file is called an outfile. Outfiles are standard text files having a .TXT extension that can be edited and printed with any editor or word processor.

Follow these steps to save the Session window.

 Step 1. Make the Session window the active window.

    Choose **Window** > **Session** or use {CTRL}M to go to the session window.

 Step 2. Save the Session window.

    Choose **File**>**Save Window As.** Select drive A: from the Save in: drop down dialog box. Place ex1_1 in the File name: text box. Choose **Save**. Observe that this file will have the extension .TXT.

Obtaining On Line Help

    Follow these steps to obtain on-line help.

 Step 1. Obtain Help.

    Choose **Help**>**Contents**, or press {F1}.

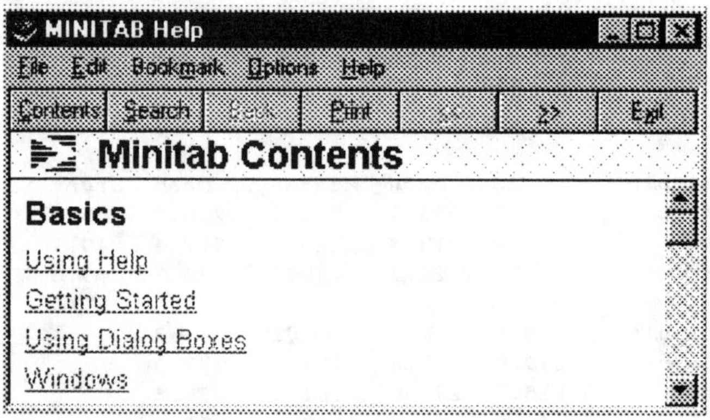

Figure 1.12

 Step 2. Select the topic.

    Double click on the text "Using Dialog Boxes", shown in Figure 1.12, (or point and {ENTER}). The Help window displays information using dialog boxes. Click on the button labeled **Back** near the top of the Help window, as shown in Figure 1.13.

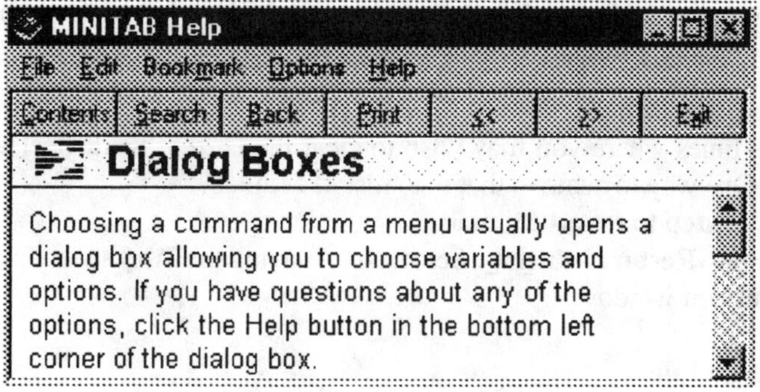

MINITAB Help

File Edit Bookmark Options Help

| Content | Search | Back | Print | << | >> | Exit |

Dialog Boxes

Choosing a command from a menu usually opens a
dialog box allowing you to choose variables and
options. If you have questions about any of the
options, click the Help button in the bottom left
corner of the dialog box.

Figure 1.13

Step 3. Using Search.

Click the button labeled **Search**.

Type Descriptive Statistics. Select "Descriptive Statistics (Stat menu)", as
shown in Figure 1.14. Click Display.

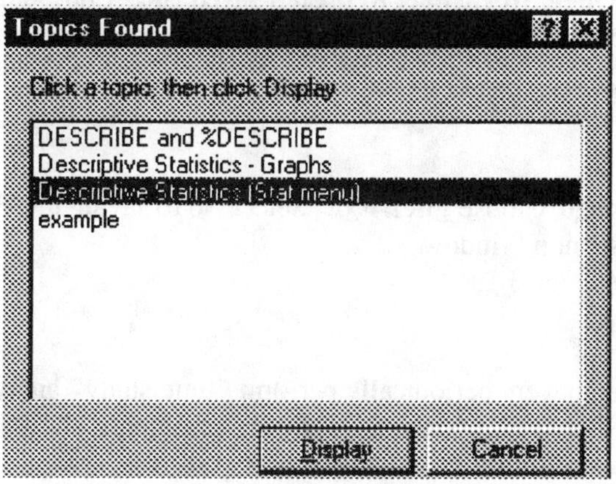

Topics Found

Click a topic, then click Display

DESCRIBE and %DESCRIBE
Descriptive Statistics - Graphs
Descriptive Statistics (Stat menu)
example

Display   Cancel

Figure 1.14

The results as shown in Figure 1.15, provides information relating to that
topic..

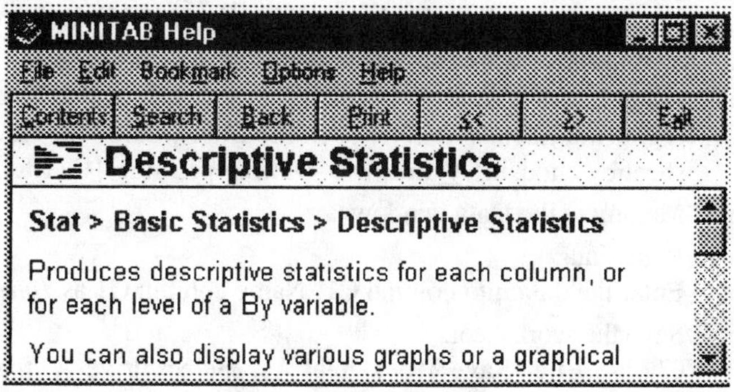

MINITAB Help

File Edit Bookmark Options Help

| Content | Search | Back | Print | << | >> | Exit |

Descriptive Statistics

Stat > Basic Statistics > Descriptive Statistics

Produces descriptive statistics for each column, or
for each level of a By variable.

You can also display various graphs or a graphical

Figure 1.15

Step 4. Exiting Help.
   To return to the Minitab session, choose the **Ex**it button from the menu bar or choose **File**>**Ex**it.

Restarting Minitab

There are times where you may wish to clear the current worksheet and Session window. One way to achieve that is to restart Minitab.
Follow this step to restart Minitab.
Choose **File**>**Restart Minitab**. Select No to saving the Worksheet and No to saving the Session window.

Retrieving a File

Besides typing data directly into the Data window there are other ways to enter data into the Minitab Data window. One of those methods involves retrieving an existing data worksheet.

Follow this step to retrieve a Minitab Worksheet.
Choose **File**>**Open Worksheet** to open a saved file. Choose drive A: from the Look **i**n: drop down dialog box. Choose the previously saved file ex1_1.mtw. Choose **O**pen.

Exiting Minitab

To exit Minitab, choose **File**>**Ex**it. Select No to saving the Worksheet and No to saving the Session Window.

Exercises

1. Industrial engineers periodically perform "time study" analyses to determine the time required to produce a single unit of output. The following measurements resulted from a recent study: 194, 198, 169, 208, 205, 224, 270, 263, 252.

   Step 1. Start Minitab.
      Choose **Start**>**Programs**. Locate and point to the Minitab program group. "Double click" on the blue Minitab for Windows icon. (If you are already in Minitab, choose **File**>**Restart Minitab** to begin with a clean worksheet.)

   Step 2. Open the Data window.
      If it is not already visible, open the Data window to view the worksheet by choosing **Window** > **Data** or pressing {CTRL}D.

   Step 3. Maximize the Data window.

   Step 4. Enter data.
      Enter the data into column C1. Name column C1 as *Time*.

   Step 5. Save the worksheet.
      Choose **File**>**Save Worksheet As**. Select drive A: from the Save **i**n: drop down dialog box. Place *1Prob1* in the File **n**ame: text box.

Choose <u>S</u>ave.

Step 6. Obtain descriptive statistics.

Choose <u>Stat</u>><u>B</u>asic Statistics><u>D</u>escriptive Statistics. Place Time in the Variables: text box. Choose <u>O</u>K.

Step 7. Make the Session window the active window.

Choose <u>W</u>indow><u>S</u>ession to go to the Session window.

Step 8. Select the correct printer.

Choose <u>F</u>ile><u>Prin</u>t Setup to select the correct printer. Choose <u>O</u>K.

Step 9. Choose <u>F</u>ile><u>P</u>rint Window to print the Session window. Choose <u>O</u>K.

Step 10. Restart Minitab.

Choose <u>F</u>ile><u>R</u>estart Minita<u>b</u>. Select No to saving the Worksheet and No to saving the Session window.

2. Time Study 2.

Step 1. Open a file.

Choose <u>F</u>ile><u>O</u>pen Worksheet. Select drive A: from the Look <u>in</u>: drop down dialog box. Place 1Prob1.mtw. Choose <u>O</u>pen.

Step 2. Enter additional data.

Enter the following times collected for a second time study in column C2: 174, 184, 152, 165, 203, 180, 182, 169, 161. Name column C2 as *Time2*.

Step 3. Save the worksheet.

Choose <u>F</u>ile><u>S</u>ave Worksheet <u>A</u>s. Select drive A: from the Save <u>in</u>: drop down dialog box. Type *1Prob2* in the File <u>n</u>ame: text box. Choose <u>S</u>ave.

Step 4. Calculate the difference between Time and Time2.

Choose <u>C</u>alc><u>C</u>al<u>c</u>ulator. Type *Difference* in the <u>S</u>tore result in variable: text box. Place Time - Time2 in the Expression: text box. Choose <u>O</u>K.

Step 5. Obtain descriptive statistics.

Choose <u>Stat</u>><u>B</u>asic Statistics><u>D</u>escriptive Statistics. Place Difference in the Variables: text box. Choose <u>O</u>K.

Step 6. Make the Session window the active window.

Choose <u>W</u>indow><u>S</u>ession to go to the Session window.

Step 7. Enter a brief description of the mean difference in between Time and Time2.

Choose E<u>d</u>itor><u>M</u>ake <u>O</u>utput Editable to enable you to enter a comment.

Step 8. Select the correct printer.

Choose <u>F</u>ile><u>Prin</u>t Setup to select the correct printer. Choose <u>O</u>K.

Step 9. Choose <u>F</u>ile><u>P</u>rint Window to print the Session window. Choose <u>O</u>K.

Chapter 2
Data Displays

Stem-and-Leaf Displays and Boxplots

2.1 Overview

A successful engineer has the capability to convert data into meaningful information, explaining the nature of the data observed and accounting for the variation within the data. One key to developing good models of the data is "listening" to what the data have to say, developing a conversation with the data. Clever plots of data provide a powerful basis for starting this conversation with data, eliciting meaningful information. Data displays can provide powerful insights relevant to the nature of the data and about the nature of the most appropriate model for the data. After reading this chapter you should be able to construct

- Stem-and-Leaf Displays.
- Boxplots.
- Histograms.
- Timeplots.

2.2 Graphic Modes

Minitab provides a number of different graph types for plotting single and multiple variables, as well as statistical control charts. There are two graphic modes: low-resolution and high-resolution. While both modes may display the same information, the high-resolution mode offers pictorial elements like lines and colors and provides a professional graphic presentation. Low resolution graphs are called character graphs are made up of normal keyboard characters. Low-resolution graphics can be viewed on any screen, printed on any printer, and stored in an outfile.

2.3 Stem-and-Leaf Displays

New Minitab Commands

1. **Graph>Character Graphs>Stem-and-Leaf** - Produces a character-based stem-and-leaf plot in the Session window. In this section, you will construct stem-and-leaf plots for large data sets.

The Basic Stem-and-Leaf Display

The stem-and-leaf display provides a means of exploring a data set where one can have an intuitive feel for the shape of the data set. Let's look at the following problem to construct the basic stem-and-leaf plot.

Example 2.1 Wall Thicknesses of Aircraft Parts

Eck Industries, Inc. (see Mee 1990) manufactures cast aluminum cylinder heads that are used for liquid-cooled aircraft engines. The wall thicknesses of the coolant jackets are critical, particularly in high-altitude applications. The thicknesses of 18 cylinder heads appear in Table 2.1.

.223	.193	.218	.201	.231	.204
.228	.223	.215	.223	.237	.226
.214	.213	.233	.224	.217	.210

Table 2.1

Follow these steps to construct the basic stem-and-leaf display:

Step 1. Enter Data.

Enter the data into column C1.

Name column C1 as *Thickness*.

Step 2. Construct the stem-and-leaf display.

Choose **Graph**>**Character Graphs**>**Stem-and-Leaf**. Place Thickness in the **V**ariables: text box. Place .01 in the **I**ncrement text box. Choose **OK**.

The Minitab Output

Character Stem-and-Leaf Display

```
Stem-and-leaf of Thicknes  N  = 18
Leaf Unit = 0.0010


    1     19 3
    3     20 14
    9     21 034578
    9     22 333468
    3     23 137
```

Figure 2.1

The section of the session window labeled Character Stem-and-Leaf Display, as shown in Figure 2.1, indicates the number of observations (N) and depth information. To the left of the stem, Minitab indicates the cumulative number of observations, counting in from the extremes. This is the depth information. The depth represents how far the observation on the right is from the appropriate end of the data set. For example, the value .204, is represented as the 4 on the .20 stem and is the third observation from the beginning of the ordered data set. The stem in which the median occurs is indicated in parentheses, and displays the frequency for that stem alone.

The objective of data displays is to engage in a conversation with the data, obtaining good information from the display about the data set. Typically, the data may be described numerically in measures of the center and measures of variability. Other questions may refer to the shape of the data set: is the shape of the data symmetric; skewed right or left; are there multiple peaks, or outliers?

The stem-and-leaf display shown in Figure 2.1 indicates that the typical wall thickness is around .21 to .22 inches. The data range from .193 to .237 inches with most of the data appearing between .210 and .230 inches. The data set appears to be symmetric with no outliers.

The Stretched Stem-and-Leaf Display
The number of stems employed in the stem-and-leaf display plays a critical part in our ability to discern patterns in a data set. If too few stems are used, then all of the leaves appear on a few stems and no discernible pattern is seen. On the other hand, if too many stems are used, the data set appears to have no pattern whatsoever. Some extensions to the basic stem-and-leaf displays address these issues.

Example 2.2 Ambient Levels of Peroxyacyl Nitrates
Williams, Grosjean, and Grosjean (1993) studied the ambient levels of peroxyacyl nitrates in Atlanta, Georgia, during the period July 22 through August 26, 1992. Peroxyacyl nitrates are eye irritants and possible skin cancer agents. The daily maxima ambient levels of a common form of peroxyacyl nitrate known as PAN (in ppb) during the period studied in 1992 are listed in Table 2.2.

0.3	1.0	0.8	1.1	1.3	1.1	2.4
2.9	1.6	1.3	0.4	0.7	1.5	1.3
2.1	1.9	1.0	1.8	1.1	2.2	1.9
0.7	0.4	1.5	0.7	2.4	2.4	2.8
2.7	1.1	1.2	1.1	1.1	0.9	1.3

Table 2.2

Follow these steps to construct the basic stem-and-leaf display:
Step 1. Enter Data.
Enter the data into column C1.
Name column C1 as *PAN*.
Step 2. Construct the stem-and-leaf display.
Choose **Graph**>**Character Graphs**>**Stem-and-Leaf**. Place PAN in the Variables: text box. Place 1 in the Increment: text box. Choose **OK**.

The Minitab Output

Character Stem-and-Leaf Display

```
Stem-and-leaf of PAN        N  = 35
Leaf Unit = 0.10


     8        0  34477789
   (19)       1  0011111123333556899
     8        2  12444789
```
Figure 2.2

The Minitab output, as shown in Figure 2.2, indicates that there are only three stems and as a consequence we cannot see the patterns in the data very well. An alternative is to stretch the stems of the data display.

Follow this step to construct the stretched stem-and-leaf display:

Choose **Graph**>**Character Graphs**>**Stem-and-Leaf**. Place PAN in the Variables: text box. Place 0.5 in the Increment: text box. Choose **OK**.

The stretched display allows us to discern patterns in this data set.

The Minitab Output

Character Stem-and-Leaf Display

```
Stem-and-leaf of PAN        N  = 35
Leaf Unit = 0.10



     3        0  344
     8        0  77789
   (13)       1  0011111123333
    14        1  556899
     8        2  12444
     3        2  789
```
Figure 2.3

The Minitab output, as shown in Figure 2.3, indicates that typical values are from 1.0 to 1.4. If the data set is skewed, it is skewed to the right (the larger values). There appear to be no outliers.

The Squeezed Stem-and-Leaf Display

If splitting the stems into two parts is not sufficient to indicate the general shape of the data, additional splitting of the stem may be appropriate. The essential concept which underlies splitting stems is that each resulting stem must have the same number of possible values. As a result, the next level of splitting is five. That is, the interval from 10 to 19 contains 10 values. Dividing by 5 units yields an increment of 2. The resulting plot is called a squeezed stem-and-leaf display. The best way to illustrate this technique is through an example.

Example 2.3 Grades on a Homework Assignment

The homework scores on the first assignment in a Honors statistics class which consisted of 11 freshmen and 9 upperclassmen follow. The course was given in the Fall semester, so the freshmen were just making the transition from high school to college.

19	16	23	22	24
25	15	19	23	23
23	17	25	23	20
18	24	17	18	18

Table 2.3

Follow these steps to construct the squeezed stem-and-leaf display:

Step 1. Enter Data.

Enter the data into column C1.

Name column C1 as *Homework*.

Step 2. Construct the stem-and-leaf display.

Choose **Graph**>**Character Graphs**>**Stem-and-Leaf**. Place Homework in the Variables: text box. Place 2 in the Increment: text box. Choose **OK**.

The Minitab Output

Character Stem-and-Leaf Display

```
Stem-and-leaf of Homework   N  = 20
Leaf Unit = 1.0


    1       1 5
    4       1 677
    9       1 88899
   10       2 0
   10       2 233333
    4       2 4455
```

Figure 2.4

The Minitab output, as shown in Figure 2.4, shows the resulting squeezed stem-and-leaf display. This plot shows two clear peaks: one at the interval containing 18 and 19 and the other peak occurring at the interval containing 22 and 23. The explanation for the two peaks may be found in a further analysis of the data set, breaking the data down by Freshmen and Upperclassmen, as shown in Table 2.4.

Freshmen				Upperclassmen		
19	16	22	24	23	25	15
19	23	17	20	23	23	25
18	18	18		23	24	17

Table 2.4

Follow these steps to construct the stem-and-leaf displays to compare two data sets:

Step 1. Enter Data.

Enter the data for both groups of students into column C1. Enter codes (Freshmen = 1, Upperclassmen = 2) for each class in column C2.

Name column C1 as *Homework*. Name column C2 as *Class*.

Step 2. Construct the stem-and-leaf display.

Choose **Graph**>**Character Graphs**>**Stem-and-Leaf**. Place HomeWork in the Variables: text box. Place a check in the By variable: checkbox. Place Class in the By variable: text box. Place 2 in the Increment: text box. Choose **OK**.

The Minitab Output

Character Stem-and-Leaf Display

```
Stem-and-leaf of Homework   Class = 1        N  = 11
Leaf Unit = 1.0

    2      1 67
   (5)     1 88899
    4      2 0
    3      2 23
    1      2 4


Stem-and-leaf of Homework   Class = 2        N  = 9
Leaf Unit = 1.0

    1      1 5
    2      1 7
    2      1
    2      2
   (4)     2 3333
    3      2 455
```

Figure 2.5

The Minitab output, as shown in Figure 2.5, indicates the resulting squeezed stem-and-leaf displays to compare two data sets. The same stem structure is used for both plots. A typical score for the freshmen was just under 20 while the upperclassmen typically scored just under 25. The freshmen scores are skewed to the right, while the upperclassmen scores are skewed to the left. Two of the upperclassmen performed poorly and appear as possible outliers.

Exercises

2.1 (EXO2O1) The following are outside diameters for the barrel of a popular felt-tip-marker.

.379	.376	.379	.379	.378
.378	.377	.378	.377	.379
.378	.377	.377	.379	.378
.377	.377	.378	.379	.378
.379	.380	.379	.378	.379
.380	.378	.379	.379	.379
.379	.380	.380	.381	.379

Follow these steps to construct stem-and-leaf displays:

Step 1. Enter Data.
Enter the data into column C1.
Name column C1 as *Diameter*.

Step 2. Construct the basic stem-and-leaf display.
Choose **Graph**>**Character Graphs**>**Stem-and-Leaf**. Place Diameter in the Variables: text box. Choose **OK**.

Step 3. Construct the stretched stem-and-leaf display.
Choose **Graph**>**Character Graphs**>**Stem-and-Leaf**. Place Diameter in the Variables: text box. Place 0.0005 in the Increment: text box. Choose **OK**.

Step 4. Construct the squeezed stem-and-leaf display.
Choose **Graph**>**Character Graphs**>**Stem-and-Leaf**. Place Diameter in the Variables: text box. Place .01 in the Increment: text box. Choose **OK**.

Looking at these three stem-and-leaf plots determine which of the three is most appropriate. Comment on your results.

2.2 (EXO2O2) Yashchin (1992) studied the thicknesses of metal wires produced in a chip-manufacturing process. Ideally, these wires should have a target thickness of 8 microns. The data, in microns, follow.

8.4	8.0	7.8	8.0	7.9	7.7	8.0	7.9	8.2	7.9
7.9	8.2	7.9	7.8	7.9	7.9	8.0	8.0	7.6	8.2
8.1	8.1	8.0	8.0	8.3	7.8	8.2	8.3	8.0	8.0
7.8	7.9	8.4	7.7	8.0	7.9	8.0	7.7	7.7	7.8
7.8	8.2	7.7	8.3	7.8	8.3	7.8	8.0	8.2	7.8

Plot an appropriate stem-and-leaf display and comment on the results.

2.3 (EXO2O3) Cryer and Ryan (1990) discuss the following chemical process data where the measurement variable is a color property.

0.67	0.63	0.76	0.66	0.69	0.71	0.72
0.71	0.72	0.72	0.83	0.87	0.76	0.79
0.74	0.81	0.76	0.77	0.68	0.68	0.74
0.68	0.69	0.75	0.80	0.81	0.86	0.86
0.79	0.78	0.77	0.77	0.80	0.76	0.67

Plot an appropriate stem-and-leaf display and comment on the results.

2.4 (EXO204) Padgett and Spurrier (1990) analyze the breaking strengths of carbon fibers used in fibrous composite materials. These fibers measure 50 mm in length and 7-8 microns in diameter. Periodically, the manufacturer selects random samples of five fibers and tests their breaking stresses. Specifications require that 99 breaking stress of at least 1.2 GPa (giga-Pascals). The breaking stresses in GPa from 20 such samples follow.

3.7	2.7	2.7	2.5	3.6	3.1	3.3	2.9	1.5	3.1
4.4	2.4	3.2	3.2	1.7	3.3	3.1	1.8	3.2	4.9
3.8	2.4	3.0	3.0	3.4	3.0	2.5	2.7	2.9	3.2
3.4	2.8	4.2	3.3	2.6	3.3	3.3	2.9	2.6	3.6
3.2	2.4	2.6	2.6	2.4	2.8	2.8	2.2	2.8	1.9
1.4	3.7	3.0	1.4	1.0	2.8	4.9	3.7	1.8	1.6
3.2	1.6	0.8	5.6	1.7	1.6	2.0	1.2	1.1	1.7
2.2	1.2	5.1	2.5	1.2	3.5	2.2	1.7	1.3	4.4
1.8	0.4	3.7	2.5	0.9	1.6	2.8	4.7	2.0	1.8
1.6	1.1	2.0	1.6	2.1	1.9	2.9	2.8	2.1	3.7

Plot an appropriate stem-and-leaf display and comment on the results.

2.5 (EXO205) The National Bureau of Standards (see Mulrow et al., 1988) tested 14 samples of biphenyl measured on a differential calorimeter calibrated with two standards in order to establish this substance's melting point. The data follow.

343.0	342.4	343.4	343.1	343.3	343.7	343.5
343.1	343.3	343.4	343.8	343.3	343.3	343.3

Plot an appropriate stem-and-leaf display and comment on the results.

2.6 (EXO206) Montgomery and Peck (1982) look at the time required to deliver cases of a popular soft drink to vending machines. The data follow and represent the time, in minutes, required by the driver to stock a machine.

16.7	11.5	12.0	14.9	13.8
18.1	8.0	17.8	79.2	21.5
40.3	21.0	13.5	19.8	24.0
29.0	15.4	19.0	9.5	35.1
17.9	52.3	18.8	19.8	10.75

Plot an appropriate stem-and-leaf display and comment on the results.
Follow these steps to construct stem-and-leaf displays:

Step 1. Enter Data.
Enter the data into column C1.
Name column C1 as *Time*.

Step 2. Construct the basic stem-and-leaf display.
Choose **Graph**>**Character Graphs**>**Stem-and-Leaf**. Place Time in the **V**ariables: text box. Choose **OK**.

Step 3. Construct the stretched stem-and-leaf display.
Choose **Graph**>**Character Graphs**>**Stem-and-Leaf**. Place Time in the **V**ariables: text box. Place 5 in the **I**ncrement: text box. Choose **OK**.

Step 4. Construct the squeezed stem-and-leaf display.
Choose **Graph**>**Character Graphs**>**Stem-and-Leaf**. Place Time in the **V**ariables: text box. Place 20 in the **I**ncrement: text box. Choose **OK**.

Looking at these three stem-and-leaf plots determine which of the three is most appropriate. Comment on your results.

2.7 (EXO207) A two-component distillation column produced these yields for 25 successive batches:

0.99	0.93	0.95	0.99	0.89
0.96	0.94	0.96	0.99	0.99
0.98	0.81	0.97	0.92	0.99
0.97	0.99	0.96	0.94	0.99
0.99	0.99	0.99	0.80	0.95

Plot an appropriate stem-and-leaf display and comment on the results.

2.8 (EXO208) Albin (1990) studied aluminum contamination in recycled PET plastic from a pilot plant operation at Rutgers University. She collected 26 samples and measured, in parts per million (ppm) the amount of aluminum contamination. The maximum acceptable level of aluminum contamination, on the average, is 220 ppm. The data follow.

291	222	125	79	145	119	244	118	182
63	30	140	101	102	87	183	60	191
119	511	120	172	70	30	90	115	

Plot an appropriate stem-and-leaf display and comment on the results.

2.9 (EXO2O9) Lucas (1985) studied the rate of accidents at a Du Pont facility over a ten year period. The following data are the number of industrial accidents per calendar quarter for the first five year period (Period I) and for the second (Period II).

	Period I				Period II		
5	5	10	8	3	4	2	0
4	5	7	3	1	3	2	2
2	8	6	9	7	7	1	4
5	6	5	10	1	2	2	1
6	3	3	10	4	4	4	4

Plot an appropriate stem-and-leaf display and comment on the results.

2.10 (EXO210) Eibl, Kess and Pukelsheim (1992) studied the impact of viscosity upon observed coating thickness produced by a paint operation. Ideally, this process should produce a coating thickness of 0.8 mm. For simplicity, they chose to study only two viscosities; "low" and "high." The data follow.

"Low" Viscosity

1.09	1.12	0.83	0.88	1.62	1.49	1.48	1.59
0.88	1.29	1.04	1.31	1.83	1.65	1.71	1.76

"High" Viscosity

1.46	1.51	1.59	1.40	0.74	0.98	0.79	0.83
2.05	2.17	2.36	2.12	1.51	1.46	1.42	1.40

Plot an appropriate stem-and-leaf display and comment on the results.

2.11 (EXO211) A major manufacturer of aircraft (see Montgomery 1991, pp 242-244) closely monitors the viscosity of an aircraft primer paint. The viscosities for two different time periods appear below.

Time Period 1

33.8	33.1	34.0	33.8	33.5
34.0	33.7	33.3	33.5	33.2
33.6	33.0	33.5	33.1	33.8

Time Period 2

33.5	33.3	33.4	33.3	34.7
34.8	34.6	35.0	34.8	34.5
34.7	34.3	34.6	34.5	35.0

Plot an appropriate stem-and-leaf display and comment on the results.

2.4   Boxplots

New Minitab Commands

1. **Graph> Boxplot** - Produces a boxplot. A default boxplot consists of a box, whiskers, and outliers. Minitab draws a line across the box at the median. In this section, you will construct a boxplot from a data set. `

    a. **Options** - Contains one options specific to Boxplot. You can transpose X and Y. Place a check in the Transpose checkbox to interchange the variables defining the vertical and horizontal axes.

The boxplot provides a quick display of some important features of the data. The boxplot "distills" the data set to its most important features and provides a formal tool for discriminating outliers during preliminary data analysis. Let's look at the following problem to construct a boxplot.

Example 2.4 Boxplot for the Wall Thicknesses of Aircraft Parts

Eck Industries, Inc. (see Mee 1990) manufactures cast aluminum cylinder heads that are used for liquid-cooled aircraft engines. The data was used in Example 2.1 and are provided once again in Table 2.5.

.223	.193	.218	.201	.231	.204
.228	.223	.215	.223	.237	.226
.214	.213	.233	.224	.217	.210

Table 2.5

Follow these steps to construct the boxplot:

Step 1. Enter Data.

   Enter the data into column C1.

   Name column C1 as *Thickness*.

Step 2. Construct the boxplot.

   Choose **Graph> Boxplot**. Place Thickness in the Y Graph Variables: Y(measurement) vs X(category) text box.

   Change the boxplot from a vertical to a horizontal boxplot.

   Choose Options. Place a check in the Transpose X and Y checkbox.

   Choose **OK**.

   Choose **OK**.

The Minitab Output

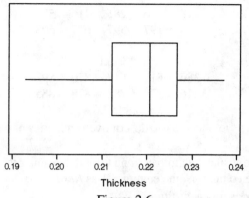

0.19   0.20   0.21   0.22   0.23   0.24

Thickness

Figure 2.6

An examination of the boxplot, as shown in Figure 2.6, indicates that a typical value is .220. The boxplot graphically depicts the position of the quartiles: Q_1, Q_2, Q_3 and indicates that 50% of the data fall between .213 (Q_1) and .226 (Q_3). The end of the left whisker indicates the minimum value (.193) and the end of the right whisker indicates the maximum value (.237) Observe that in this case .193 is not recognized as an outlier by Minitab (outliers will be identified by a *).

Parallel Boxplots

New Minitab Commands

1. **Stat>Basic Statistics>Descriptive Statistics** - Produces descriptive statistics (N, Mean, Median, Standard Deviation, etc.) for each variable or column. In this section, you will produce descriptive statistics for a small data set.

Parallel or side-by-side boxplots provide a means of comparing two or more data sets simultaneously. The following example illustrates this technique.

Example 2.5 Sensory Modalities

Galinsky et al. (1993) studied the impact of sensory modalities (either aural or visual) upon people's ability to monitor a specific display for critical events to which the observer must respond. Such tasks are critical components of such jobs as air traffic control, industrial quality control, robotic manufacturing operations, and nuclear power plant monitoring. One aspect of their study focused upon the difference in response between aural and visual stimuli. In particular, they monitored the motor activity of the subject's dominant wrist as a measure of "restlessness" or "fidgeting"; the greater the activity, the more restless the subject. Galinsky and her colleagues recorded the number of wrist movements over ten minute periods of time. The data follow.

Auditory

418	236	281	416	578
329	197	397	677	698

Visual

386	517	617	870	892
416	574	782	838	885

Table 2.6

Follow these steps to obtain basic descriptive summaries of the data:

Step 1. Enter Data.

Enter the Visual data into column C1and then enter the Auditory data in the same column. Name column C1 as *Restless*.

Enter codes for each stimuli.

Enter ten 1's as the code for the Visual data followed by ten 2's as the code for the Auditory data into column C2. Name column C2 as *Stimuli*.

Step 2. Obtain basic statistics.

Choose **Stat**>**Basic Statistics**>**Descriptive Statistics**. Enter Restless in the **V**ariables: text box. Place a check in the **B**y variable: checkbox. Place Code in the **B**y variable: text box. Choose **OK**.

The Minitab Output

Descriptive Statistics

Variable	Stimuli	N	Mean	Median	Tr Mean	StDev	SE Mean
Restless	1	10	422.7	406.5	416.5	176.4	55.8
	2	10	677.7	699.5	687.4	199.0	62.9

Variable	Stimuli	Min	Max	Q1	Q3
Restless	1	197.0	698.0	269.7	602.8
	2	386.0	892.0	491.7	873.7

Figure 2.7

Follow these steps to construct the parallel boxplots:

Step 1. Enter Data.

Enter the Visual data into column C1and then enter the Auditory data in the same column. Name column C1 as *Restless*.

Enter ten 1's as the code for the Visual data followed by ten 2's as the code for the Auditory data into column C2. Name column C2 as *Stimuli*.

Step 2. Construct the boxplot.

Choose **Graph**> **Boxplot**. Place Restless in the Y **G**raph Variables: Y(measurement) vs X(category) text box. Place Stimuli in the X **G**raph Variables: Y(measurement) vs X(category) text box.

Change the boxplot from a vertical to a horizontal boxplot.

Choose O**p**tions. Place a check in the **T**ranspose X and Y checkbox. Choose **OK**.

Enter a Title.

Choose <u>A</u>nnotation><u>T</u>itle. Place an appropriate title (*Aural vs. Visual Activities for Restlessness*) on line 1. Place the title *Visual = 1 Aural = 2* on line 2. Choose **OK**.

Choose **OK**.

The Minitab Output

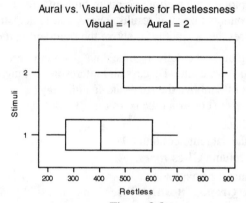

Aural vs. Visual Activities for Restlessness
Visual = 1   Aural = 2

Figure 2.8

There appear to be no apparent outliers in either data set, as indicated by the parallel box plots shown in Figure 2.8. Visual activities appear to lead to more wrist movement, thus indicating more restlessness, than auditory activities.

Exercises

2.12 Construct a boxplot for the felt-tip marker data given in Exercise 2.1 of the text and comment on your results.

Follow these steps to construct the boxplot:

Step 1. Enter Data.

Enter the data into column C1.

Name column C1 as *Diameter*.

Step 2. Construct the boxplot.

Choose **<u>G</u>raph**> **<u>B</u>oxplot**. Place Diameter in the Y <u>G</u>raph Variables: Y(measurement) vs X(category) text box.

Change the boxplot from a vertical to a horizontal boxplot.

Choose <u>O</u>ptions. Place a check in the <u>T</u>ranspose X and Y checkbox. Choose **OK**.

Enter a Title.

Choose <u>A</u>nnotation><u>T</u>itle. Place an appropriate title on line(s) 1 (and 2). Choose **OK**.

Choose **OK**.

2.13 Construct a boxplot for the thicknesses of metal wires given in Exercise 2.2 of the text and comment on your results. Discuss and interesting features. What insights does the boxplot offer above the stem-and-leaf display?

2.14 Construct a boxplot of the color property data given in Exercise 2.3 of the text and comment on your results. Discuss and interesting features. What insights does the boxplot offer above the stem-and-leaf display?

2.15 Construct a boxplot for the breaking strength of carbon fibers data given in Exercise 2.4 of the text and comment on your results. Discuss and interesting features. What insights does the boxplot offer above the stem-and-leaf display?

2.16 Construct a boxplot for the melting points of biphenyl given in Exercise 2.5 of the text and comment on your results. Discuss and interesting features. What insights does the boxplot offer above the stem-and-leaf display?

2.17 Construct a boxplot for the delivery time data given in Exercise 2.6 of the text and comment on your results. Discuss and interesting features. What insights does the boxplot offer above the stem-and-leaf display?
   Follow these steps to construct the boxplot:

   Step 1. Enter Data.
            Enter the data into column C1.
            Name column C1 as *Times*.

   Step 2. Construct the boxplot.
            Choose **Graph**> **Boxplot**. Place Times in the Y Graph Variables: Y(measurement) vs X(category) text box.

            Change the boxplot from a vertical to a horizontal boxplot.
            Choose Options. Place a check in the Transpose X and Y checkbox.
            Choose **OK**.

            Enter a Title.
            Choose Annotation>Title. Place an appropriate title on line(s) 1 (and 2). Choose **OK**.

            Choose **OK**.

2.18 Construct a boxplot for the yields from the two-component distillation process given in Exercise 2.7 and comment on your results. Discuss and interesting features. What insights does the boxplot offer above the stem-and-leaf display?

2.19 Construct a boxplot for the amounts of aluminum contamination given in Exercise 2.8 and comment on your results. Discuss and interesting features. What insights does the boxplot offer above the stem-and-leaf display?

2.20 Construct parallel boxplots for the accident data given in Exercise 2.9 and comment on your results. Discuss and interesting features. What insights does the boxplot offer above the stem-and-leaf display?

2.21 Construct parallel boxplots for the coating thickness given in Exercise 2.10 and comment on your results. Discuss and interesting features. What insights does the boxplot offer above the stem-and-leaf display?

2.22 Construct parallel boxplots for the paint viscosities given in Exercise 2.11 and comment on your results. Discuss and interesting features. What insights does the boxplot offer above the stem-and-leaf display?

2.23 Construct parallel boxplots for the yields given in Example 2.12 and comment on your results. Discuss and interesting features. What insights does the boxplot offer above the stem-and-leaf display?

2.24 (EXO226) George et al. (1994) studied three different thermoplastic starches: waxy maize; native corn; and high amylose corn. They injection molded 2 mm thick test specimens at a melt temperature of 350 degrees F, an injection speed of 3 in/sec., a hold pressure of 5000 psi, a hold time of 5 sec., and a cool time of 10 sec. For each sample, they recorded the minimum injection pressure which measures the processability of the thermoplastic starch. A lower minimum injection pressure indicates easier processability. The pressures, in thousands of psi, for each thermoplastic starch follow.

Waxy Maize

| 13.0 | 9.0 | 10.0 | 10.0 |
| 10.0 | 6.0 | 7.0 | 7.0 |

Native Corn

| 22.5 | 18.0 | 9.0 | 9.0 |
| 9.0 | 6.0 | 10.0 | 10.0 |

High Amylose Corn

| 15.0 | 13.0 | 18.0 | 14.5 |
| 12.0 | 11.0 | 8.9 | 8.0 |

Construct appropriate parallel boxplots for these data and comment on your results.

Stem-and-Leaf Displays, Boxplots, Histograms, and Time Plots

In addition to stem-and-leaf displays and boxplots, Minitab can produce other graphical displays such as histograms and time plots. These latter two topics, histograms and time plots are briefly examined in this section.

Example 2.6 Times Between Industrial Accidents

Lucas (1985) studied the times between accidents for a ten year period at a DuPont facility. DuPont historically has strongly emphasized the importance of safety in its operations and has always striven to reduce the number of accidents over time. During this period, 178 accidents occurred.. Table 2.5 in the text gives the data, which are the number of days since the previous accident.

Stem-and-Leaf Displays and Boxplots

Follow these steps to construct the basic stem-and-leaf display:

Step 1. Enter Data or retrieve the file.
        Enter the data into column C1.
        Name column C1 as *Times*.

Step 2. Construct the stem-and-leaf display.
        Choose **Graph**>**Character Graphs**>**Stem-and-Leaf**. Place Times in
        the **V**ariables: text box. Place 5 in the **I**ncrement: text box. Choose **OK**.

The Minitab Output

Character Stem-and-Leaf Display

```
Stem-and-leaf of Times      N  = 177
Leaf Unit = 1.0

   41     0 0000000000111111122222222233333333333344444
   69     0 5566666777777777778888999999
  (25)    1 00011111122233333444444444
   83     1 555556688888999
   68     2 00000112222233344
   51     2 566777889
   42     3 00123444
   34     3 556666899
   25     4 0334
   21     4 689
   18     5 1333
    .     . ...
    1    17 3
```

Figure 2.9

The shape of the stem-and-leaf plot indicates that the times between accidents is extremely skewed to the right. The time between accidents cannot be less than zero, so we encounter a natural lower bound for these times. On the other hand, it is entirely possible to go significant periods of time without an accident, so extreme values tend to be large - hence, the skew in our data. The outliers indicated in the stem-and-leaf plot are also reflected in the boxplot shown in Figure 2.10.

Times Between Accidents

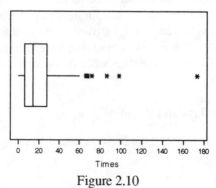

Figure 2.10

Histograms

New Minitab Commands

1. **Graph**>**Histogram** - Produces a histogram. Separates the data into intervals on the x-axis, and draws a bar for each interval whose height, by default, is the number of observations (or frequency) in the interval. In this section, you will construct a histogram for a data set.

   a. **Options** - Contains options specific to Histogram. In this section, you will construct histograms with the frequency scale on the y-axis by darkening the option button for the Frequency Type of histogram. You will also control the divisions on the x-axis by darkening the Midpoint/cutpoint positions: option button.

   b. **Annotation**>**Title** - Creates titles for the histogram, placed at the top of the figure region, above the data region. You can create as many titles as you want, specifying different attributes for each title in the corresponding attributes rows (font, color, size, and so on). In this section, you will place a first title on the histogram.

Histograms provide another opportunity to see the shape of the data. Frequency histograms reflect the number of observations that fall into a number of intervals over the range of the data. If you reexamine the data in Example 2.6, Times Between Accidents, the frequency histogram should reflect a definite right-tailed pattern.

Follow these steps to construct a histogram of the times between accidents data.

Step 1. Enter Data or retrieve the file.
Enter the data into column C1.
Name column C1 as *Times*.

Step 2. Choose intervals for the histogram.
Since the observations occur approximately over the interval from 0 to 100, let us choose intervals 10 units wide. In the next empty column, C5, if you saved the worksheet, place the cutpoints 10, 20, 30, 40, 50, 60, 70, 80, 90, 100. Name column C5 as *Cutpoints*.

Step 3. Construct the frequency histogram.
Choose **Graph**>**Histogram**. Place Times in the X Graph variables: text box.
Select the frequency histogram option and designate the cutpoints.
Choose Options. Darken the Frequency Type of Histogram option button.
Darken the Cutpoint Types of intervals option button. Darken the Midpoint/cutpoint positions: option button. Place Cutpoints in the Midpoint/cutpoint positions: text box. Choose **OK**.
Enter a Title.
Choose **Annotation**>**Title**. Place the title: *Times Between Accidents* in

29

the first line of the Title text box. Choose **OK**.
Choose **OK**.

The Minitab Output

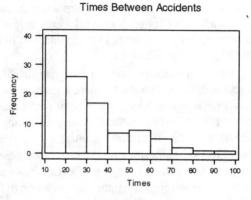

Figure 2.11

The shape of the frequency histogram, as shown in Figure 2.11, for the times between accidents clearly indicates a definite right-tailed histogram.

Time Plots

New Minitab Commands

1. **Graph>Time Series Plot -** Produces a time series plot, with time on the x-axis and the specified column on the y-axis. By default, Minitab displays symbols for each point and joins the points with a line. In this section, you will construct a time series plot for a data set.

In many cases, we observe data over time, like the times between accidents. In cases such as this, we should plot the data over time and look for trends. Time plots place the characteristic of interest on the y-axis, while the actual time or the sequence in which the data is observed is plotted on the x-axis. Some quality engineers call this plot a run chart.

Follow these steps to construct a time plot of the times between accidents data.

Step 1. Enter Data or retrieve the file.
　　　　Enter the data into column C1.
　　　　Name column C1 as *Times*.

Step 2. Construct the time plot.
　　　　Choose **Graph>Time Series Plot**. Place Times in the Y Graph
　　　　variables: text box.
　　　　Enter a title.

Choose <u>A</u>nnotation><u>T</u>itle. Place the title: *Times Between Accidents* in the first line of the Title text box. Choose <u>OK</u>. Choose <u>OK</u>.

The Minitab Output

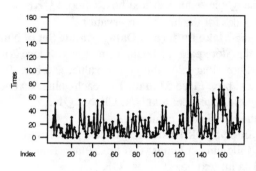

Figure 2.12

On careful inspection of the time plot, as shown in Figure 2.12, we see that the times seem to be longer during the last several years of the study.

Parallel Boxplots

New Minitab Commands

1. <u>C</u>alc><u>M</u>ake <u>P</u>atterned Data><u>S</u>imple Set of Numbers - Provides an easy way to fill a column with numbers that follow a pattern, such as the numbers 1 through 100, or five sets of 1, 2, and 3. This is very useful for entering a large number of patterned numbers. With this command, you can specify a pattern of equally spaced numbers, such as 10 20 30. In this section, you will enter a simple set of numbers.

2. <u>M</u>anip > S<u>t</u>ack/Unstack > <u>S</u>tack - Stacks a list of columns, one on top of the other, to form a new, longer column, and optionally stores a second column containing subscript (or group) values. In this section, you will stack codes for time.

In this example, Lucas points out that the accident rate seems to have dropped over the last five years as opposed to the first five. The first 120 times come from the first five years, and the last 57 times come from the last five years of the study. To create parallel boxplots, enter a 1 to represent the first five years; the value 2 represents the second five years.

Follow these steps to construct the parallel boxplots:

Step 1. Enter Codes for Time.

Enter the codes for the first 120 observations.

Chose **Calc**>**Make Patterned Data**>**Simple Set of Numbers**. Place *Time1* in the Store patterned data in: text box.. Place 1 in the From first value: text box. Place 1 in the To last value: text box. Place 1 in the In steps of: text box. Place 120 in the List each value: text box. Place 1 in the List the whole sequence: text box. Choose **OK**.

Enter the codes for the last 57 observations.

Chose **Calc**>**Make Patterned Data**>**Simple Set of Numbers**. Place *Time2* in the Store patterned data in: text box.. Place 2 in the From first value: text box. Place 2 in the To last value: text box. Place 1 in the In steps of: text box. Place 57 in the List each value: text box. Place 1 in the List the whole sequence: text box. Choose **OK**.

Stack both codes in one column.

Chose **Manip**>**Stack/Unstack**>**Stack**. Place Time1 and Time2 in the Stack the following columns: text box. Place TimeCode in the Store the stacked data in: text box. Choose **OK**.

Step 2. Construct the boxplot.

Choose **Graph**> **Boxplot**. Place Times in the Y Graph Variables: Y(measurement) vs X(category) text box. Place TimeCode in the X Graph Variables: Y(measurement) vs X(category) text box.

Change the boxplot from a vertical to a horizontal boxplot.

Choose Options. Place a check in the Transpose X and Y checkbox. Choose **OK**.

Enter a title.

Choose Annotation>Title. Place an appropriate title on line(s) 1 (and 2). Choose **OK**.

Choose **OK**.

The Minitab Output

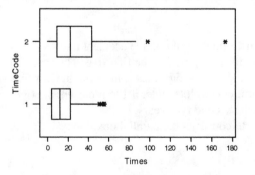

Times Between Accidents
for the First Five Years vs the Second Five Years

Figure 2.13

The boxplot, as shown in Figure 2.13, clearly shows that the times between accidents during the second five years do tend to be longer than those in the first five years. Obviously, the plant's safety efforts have born fruit.

Exercises

2.25 Use appropriate software to construct a stem-and-leaf display, a boxplot, and a histogram for the felt-tip marker data given in Exercise 2.1 of the text. Give a thorough discussion of your results, including the relative advantages and disadvantages of each display.

2.26 Use appropriate software to construct a stem-and-leaf display, a boxplot, a histogram, and a time plot for the thicknesses of metal wires given in Exercise 2.2 of the text. How well do these data meet the target of 8.0 microns? Give a thorough discussion of your results, including the relative advantages and disadvantages of each display.

2.27 Use appropriate software to construct a stem-and-leaf display, a boxplot, a histogram, and a time plot (the data are in order) for the color property data given in Exercise 2.3 of the text. Give a thorough discussion of your results, including the relative advantages and disadvantages of each display.

2.28 Use appropriate software to construct a stem-and-leaf display, a boxplot, a histogram, and a time plot for the breaking strength of carbon fibers data given in Exercise 2.4 of the text. Give a thorough discussion of your results, including the relative advantages and disadvantages of each display.

2.29 Use appropriate software to construct a stem-and-leaf display, a boxplot, and a histogram for the melting points of biphenyl given in Exercise 2.5 of the text. Give a thorough discussion of your results, including the relative advantages and disadvantages of each display.

2.30 Use appropriate software to construct a stem-and-leaf display, a boxplot, and a histogram for the delivery time data given in Exercise 2.6 of the text. Give a thorough discussion of your results, including the relative advantages and disadvantages of each display.

2.31 Use appropriate software to construct a stem-and-leaf display, a boxplot, a histogram, and a time plot for the yields given in Exercise 2.7 of the text. Give a thorough discussion of your results, including the relative advantages and disadvantages of each display.

2.32 Use appropriate software to construct a stem-and-leaf display, a boxplot, and a histogram for the amounts of aluminum contamination given in Exercise 2.8 of the text. Give a thorough discussion of your results, including the relative advantages and disadvantages of each display.

2.33 Use appropriate software to construct parallel boxplots for the accident data given in Exercise 2.9 of the text. Give a thorough discussion of your results.

2.34 Use appropriate software to construct parallel boxplots for the coating thickness given in Exercise 2.10 of the text. Give a thorough discussion of your results.

2.35 Use appropriate software to construct parallel boxplots for the paint viscosities given in Exercise 2.11 of the text. Give a thorough discussion of your results.

2.36 Use appropriate software to construct parallel boxplots for the homework scores given in Example 2.4 of the text. Give a thorough discussion of your results.

2.37 Use appropriate software to construct parallel boxplots for the minimum injection pressures given in Exercise 2.26 of the text. Give a thorough discussion of your results.

2.38 (EXO240) Snee (1983) examined the thickness of paint can ears. Periodically, the manufacturer took random samples of five cans each and measured the thickness of the ears. The data, in units of .001 inches, follow.

29	36	39	34	34	29	29	28	32	31
34	34	39	38	37	35	37	33	38	41
30	29	31	38	29	34	31	37	39	36
30	35	33	40	36	28	28	31	34	30
32	36	38	38	35	35	30	37	35	31
35	30	35	38	35	38	34	35	35	31
34	35	33	30	34	40	35	34	33	35
34	35	38	35	30	35	30	35	29	37
40	31	38	35	31	35	36	30	33	32
35	34	35	30	36	35	35	3	38	36
32	36	36	32	36	36	37	32	34	34
29	34	33	37	35	36	36	35	37	37
36	30	35	33	31	35	30	29	38	35
35	36	30	34	36	35	30	36	29	35
38	36	35	31	31	30	34	40	28	30

Follow these steps to construct the basic stem-and-leaf display:

Step 1. Enter Data.
Enter the data into column C1.
Name column C1 as *Ears*.

Step 2. Construct the stem-and-leaf display.
Choose **Graph**>**Character Graphs**>**Stem-and-Leaf**. Place Ears in the <u>V</u>ariables: text box. Choose **OK**. Construct the stem-and-leaf display once more placing an appropriate value in the <u>I</u>ncrement: text box. Choose <u>O</u>K.

Follow these steps to construct the boxplot:

Step 1. Construct the boxplot.
Choose **Graph**> **Boxplot**. Place Ears in the Y <u>G</u>raph Variables: Y(measurement) vs X(category) text box.

Step 2. Change the boxplot from a vertical to a horizontal boxplot.
Choose Options. Place a check in the <u>T</u>ranspose X and Y checkbox. Choose <u>O</u>K.

Enter a title.
Choose <u>A</u>nnotation><u>T</u>itle. Place an appropriate title on line(s) 1

(and 2). Choose **OK**.

Choose **OK**.

Follow these steps to construct the histogram:

Step 1. Construct the histogram.

Choose **Graph**>**Histogram**. Place Ears in the X Graph variables: text box.

Enter a title.
Choose Annotation>Title. Place an appropriate title in the first line of the Title text box. Choose **OK**.

Choose **OK**.

2.39 (EXO241) Nelson (1989) examined the cold cranking power of car batteries. This study used five different models and measured how many seconds an individual battery provided its rated amperage without falling below 7.2 volts at $0°F$. The results follow.

Model 1	Model 2	Model 3	Model 4	Model 5
41	42	27	48	28
43	43	26	45	32
42	46	28	51	37
46	38	27	46	25

Follow these steps to construct the parallel boxplots:

Step 1. Enter Data.

Enter the data for all models into column C1, beginning with Model 1, followed by Model 2, etc. Name column C1 as *Power*.

Step 2. Enter Codes for Models.

Choose **Calc**>**Make Patterned Data**>**Simple Set of Numbers**.
Place *Models* in the Store patterned data in: text box.. Place 1 in the From first value: text box. Place 5 in the To last value: text box. Place 1 in the In steps of: text box. Place 4 in the List each value: text box. Place 1 in the List the whole sequence: text box. Choose **OK**.

Step 3. Construct the boxplots.

Choose **Graph**> **Boxplot**. Place Power in the Y Graph Variables: Y(measurement) vs X(category) text box. Place Models in the X Graph Variables: Y(measurement) vs X(category) text box.

Change the boxplot from a vertical to a horizontal boxplot.
Choose Options. Place a check in the Transpose X and Y checkbox. Choose **OK**.

Enter a title.
Choose Annotation>Title. Place an appropriate title on line(s) 1 (and 2). Choose **OK**.

Choose **OK**.

Chapter 3
Modeling Random Behavior

3.1   Overview

Basics of Probability

Real data exhibit variability and that variability means uncertainty.. Statistics uses probability to model the random behavior of that real data as well as to quantify the uncertainty when inferences are made about a population of interest. Probability experiments via simulation may be performed a large number of times to understand underlying concepts of probability. Random samples with given probability distributions can be generated to examine and illustrate those probability distributions.

The frequentist interpretation of probability holds that the probability of an event is the number of times, f, that the event may occur, divided by the total number, N, of possible outcomes in the experiment. Thus, the probability of an event is $\frac{f}{N}$. Minitab may be used to perform the calculations required in applying the frequentist approach to probability.

After reading this chapter you should be able to

- Construct the Cumulative Distribution Function for a Discrete Probability Distribution.

- Determine the Expected Values for a Discrete Probability Distribution.

- Find the Probability Density Function for a Binomial Probability Distribution.

- Obtain the Cumulative Distribution Function for a(n)
  - Poisson Distribution.
  - Exponential Distribution.

- Obtain Areas Under the Normal Curve.

- Simulate Sampling Distributions of Sample Means.

- Construct Normal Probability Plots.

- Perform One Sample t-tests.

3.2   Random Variables and Distributions

New Minitab Commands

1. <u>C</u>alc><u>P</u>robability <u>D</u>istributions><u>D</u>iscrete Allows you to calculate probabilities, cumulative probabilities, and inverse cumulative probabilities for a discrete distribution. In this section, you will enter a discrete probability distribution in the Data window, using this command to calculate cumulative probabilities.

Random variables and their distributions are the basis for modeling and describing populations. A random variable is, very simply, a number that is assigned to the outcome of an experiment.

Example 3.1 Titanium Dioxide Production Lines

A major manufacture of titanium dioxide, the white pigment in paint, uses two separate production lines (A and B). The random variable, Y, represents the number of lines running at any one time. The probability distribution of Y is as follows.

y_i	$P(Y=y_i)$
0	0.01
1	0.28
2	0.71

Follow these steps to construct the cumulative distribution function of Y:

Step 1. Enter Data.

Enter the data under y_i into column C1. Enter the probabilities under $P(Y=y_i)$ into column C1.

Name column C1 as y and column C2 as *P(y)*.

Step 2. Construct the cumulative distribution.

Choose **Calc**>**Probability Distributions**>**Discrete**. Darken the **C**umulative Probability option button. Place y in the **V**alues in: text box and P(y) in the Pro**b**abilities in: text box. Place y in the Input column: text box and CumProb in the Optional storage: text box. Choose **OK**.

The Minitab Output

y	P(y)	CumProb
0	0.01	0.01
1	0.28	0.29
2	0.71	1.00

Figure 3.1

The Minitab output in the Data window, as shown in Figure 3.1, indicates the resulting cumulative distribution function. The table indicates that the probability of one or fewer lines running at any one time is 0.29 and that the probability of two or fewer lines running at any one time is 1.00.

3.3   Discrete Random Variables

New Minitab Commands

1. **Calc**>**Calculator** - Performs arithmetic using an algebraic expression. You can use arithmetic operations, comparison operations, logical operations, functions, and column operations. In this section, you will use this calculator to calculate the mean, variance and standard deviation when give a discrete probability distribution.

2. **Calc>Random Data>Discrete** - Generates random data from a discrete distribution. In this section, you will enter a discrete distribution (table) into two columns of Minitab. You will then use this command, to generate random data, in conjunction with other commands in order to compare the theoretical values to the sampling distribution of the variable.

A discrete random variable is one that can assume at most a countable number of values. The number of separate production lines in the Titanium Dioxide Production Lines example constitute a discrete random variable.

Expected Values
As you see, the cumulative distribution of a discrete random variable may be determined using Minitab. Minitab may also be used to determine the theoretical mean, variance and standard deviation of a discrete probability distribution.

Expected Values of Discrete Random Variables
Follow these steps to determine the theoretical mean, variance and standard deviation for the number of lines running at any one time.

Step 1. Find the theoretical mean, using the equation $\sum y \cdot p(y)$.
Choose **Calc>Calculator**. Type *mu* in the Store result in variable: text box. Select Statistics from the Functions: drop down dialog box. Select Sum from the Functions: list box. Select y from the variables list box. Click * on the calculator. Select P(y) from the variables list box. The Expression: text box should now contain Sum(y * 'P(y)'). Choose **OK**.

Step 2. Find the theoretical variance, using the equation $\sum y^2 \cdot p(y) - \mu^2$.
Choose **Calc>Calculator**. Enter Variance in the Store result in variable: text box. Select Statistics from the Functions: drop down dialog box. Select Sum from the Functions: list box. Place SUM(y**2 * 'P(y)') - mu**2 in the Expression: text box. Choose **OK**.

Step 3. Find the theoretical standard deviation by taking the square root of the variance.
Choose **Calc>Calculator**. Enter StDev in the Store result in variable: text box. Select Arithmetic from the Functions: drop down dialog box. Select Square root from the Functions: list box. Place SQRT(Variance) in the Expression: text box. Choose **OK**.

The Minitab Output

y	P(y)	CumProb	mu	Variance	StDev
0	0.01	0.01	1.7	0.23	0.479583
1	0.28	0.29			
2	0.71	1.00			

Figure 3.2

The Minitab output in the Data window, as shown in Figure 3.2, indicates the mean, μ, of the discrete distribution, 1.7, the variance, σ^2, 0.23 and the standard deviation, σ, 0.47958.

Follow these steps to compare the theoretical values to the sampling distribution.

Step 1. Generate random data.

Choose **Calc**>**Random Data**>**Discrete**. Place 1000 in the Generate rows of data text box. Place Data in the Store in column(s): text box. Place y in the Values in: text box. Place P(y) in the Probabilities in: text box. Choose **OK**.

Step 2. Describe the data.

Choose **Stat**>**Basic Statistics**>**Descriptive Statistics**. Place Data in the Variables: text box. Choose **OK**.

The Minitab Output

Descriptive Statistics

Variable	N	Mean	Median	Tr Mean	StDev	SE Mean
Data	1000	1.6640	2.0000	1.7078	0.5190	0.0164

Variable	Min	Max	Q1	Q3
Data	0.0000	2.0000	1.0000	2.0000

Figure 3.3

The Minitab output, as shown in Figure 3.3, indicates the sample statistics: the sample mean, $\overline{x} = 1.6640$ as well as the sample standard deviation, $s = 0.5190$. Compare these sample statistics to the theoretical values of $\mu = 1.7$ and $\sigma = .47958$. Your values will be different since the data is random data.

The Binomial Distribution

New Minitab Commands

1. **Calc**>>**Probability Distributions**> **Binomial** - Allows you to calculate the probability densities, cumulative probabilities, and inverse cumulative probabilities for a binomial distribution.

   a. **Probability density:** In this section, you will darken the Probability density: option button in order to determine the probability of success for all values from the binomial probability distribution.

   b. **Cumulative probability:** In this section, you will darken the Cumulative probability: option button in order to determine the cumulative probability of success for all values from the binomial probability distribution.

2. **Calc**>**Random Data**>**Binomial** - Generates random data from a binomial distribution. In this section, you will enter the number of trials and the probability of success into appropriate text boxes to generate random data. You will use

this command in conjunction with other commands in order to compare the theoretical values to the sampling distribution.

We must often model the random behavior of data which results in one of two possible outcomes, typically labeled as "success" or "failure." The binomial distribution is frequently used as the probability model.

The binomial probability experiment satisfies these conditions:

1. The experiment consists of n identical trials.

2. Each trial results in one of two outcomes: "success" or "failure."

3. The probability of a success on a single trial is p and remains the same from trial to trial. The probability of "failure" is $q = 1 - p$.

4. The trials are independent.

5. We are interested in Y, where Y represents the total number of successes among the n trials.

The probability function for a binomial random variable is

$$p(y) = \begin{cases} \binom{n}{y} \cdot p^y \cdot q^{n-y} = \frac{n!}{y!(n-y)!} \cdot p^y \cdot q^{n-y} & \text{for } y = 0, 1, 2, ..., n \\ 0 & \text{otherwise.} \end{cases}$$

Example 3.2 Non-Conforming Brick

Marcucci (1985) reports on a brick manufacturing process which classifies the product as either

   a. suitable for all purposes (standard),

   b. structurally sound but not suitable for all uses, or

   c. unacceptable for use (cull).

The latter two categories may be viewed as not meeting the standard or as non-conforming. Historically, this process produced 5% non-conforming brick. Under normal circumstances, this facility makes 25 bricks per hour.

Follow these steps to determine the probability of producing exactly two non-conforming bricks in the next hour.

Step 1. Enter Patterned Data.

   Choose **Calc**>**Make Patterned Data**>**Simple Set of Numbers**. Type *Bricks* in the Store patterned data in: text box. Place 0 in the From first value: text box and 25 in the To last value: text box. Choose **OK**.

Step 2. Calculate Probability Densities.

   Choose **Calc**>**Probability Distributions**>**Binomial**. Darken the Probability option button. Place 25 in the Number of trials: text box. Place .05 in the Probability of success: text box. Place Bricks in the Input column: text box. Choose **OK**.

The Minitab Output

Probability Density Function

Binomial with n = 25 and p = 0.0500000

x	P(X = x)
0.00	0.2774
1.00	0.3650
2.00	0.2305
3.00	0.0930
4.00	0.0269
5.00	0.0060
6.00	0.0010
7.00	0.0001
. . .	. . .
25.00	0.0000

Figure 3.4

The Minitab output, as shown in Figure 3.4, indicates the resulting probability density function. The table indicates that the probability of exactly two non-conforming bricks is 0.2305.

Follow this step to construct the cumulative distribution function of Y:

Choose **Calc**>**Probability Distributions**>**Binomial**. Darken the **C**umulative probability option button. Place 25 in the N**u**mber of trials: text box. Place .05 in the Pro**b**ability of success: text box. Place Bricks in the Input co**l**umn: text box. Choose **OK**.

The Minitab Output

Cumulative Distribution Function

Binomial with n = 25 and p = 0.0500000

x	P(X <= x)
0.00	0.2774
1.00	0.6424
2.00	0.8729
3.00	0.9659
4.00	0.9928
5.00	0.9988
6.00	0.9998
. .	. . .
25.00	1.0000

Figure 3.5

The Minitab output, shown in Figure 3.5, indicates the resulting cumulative distribution function. The table indicates that the probability of 0 (or fewer) non-conforming bricks is 0.2774. The probability of at least one non-conforming brick in the next hour's production is $1 - P(Y < 1) = 1 - P(Y = 0) = 1 - 0.2774 = .7226$. Consequently, at least one non-conforming brick an hour is expected to occur over 70% of the time.

The mean, variance, and standard deviation of the binomial probability distribution are: $\mu = np$, $\sigma^2 = npq$, and $\sigma = \sqrt{npq}$. In this example,

$$\mu = np = 25 * (0.05) = 1.25 \,,$$
$$\sigma^2 = npq = 25 * (0.05) * (0.95) = 1.1875 \,, \text{ and}$$
$$\sigma = \sqrt{npq} = \sqrt{1.1875} = 1.0897.$$

Follow these steps to compare the theoretical values to the sampling distribution.

Step 1. Generate random data.

Choose **Calc**>**Random Data**>**Binomial**. Place 1000 in the Generate rows of data text box. Type *Data* in the Store in column(s): text box. Place 25 in the Number of trials: text box. Place .05 in the Probability of success: text box. Choose **OK**.

Step 2. Describe the data.

Choose **Stat**>**Basic Statistics**>**Descriptive Statistics**. Place Data in the Variables: text box. Choose **OK**.

The Minitab Output

Descriptive Statistics

Variable	N	Mean	Median	Tr Mean	StDev	SE Mean
Data	1000	1.2930	1.0000	1.2300	1.0790	0.0341

Variable	Min	Max	Q1	Q3
Data	0.0000	6.0000	0.0000	2.0000

Figure 3.6

The Minitab output, shown in Figure 3.6, indicates that the mean of the sample data is 1.2930 and the standard deviation is 1.0790. Compare these values for the sampling distribution to the theoretical mean and standard deviation for the binomial probability distribution: 1.25 and 1.0897, respectively. Your output may differ since this data is random data.

The Poisson Distribution

New Minitab Commands

1. **Calc**>**Probability Distributions**> **Poisson** - Allows you to calculate the probability densities, cumulative probabilities, and inverse cumulative probabilities for a Poisson distribution.

2. **Calc**>**Random Data**>**Poisson** - Generates random data from a Poisson distribution. In this section, you will enter the mean into an appropriate text box to generate random data. You will use this command in conjunction with other commands in order to compare the theoretical values to the sampling distribution.

The binomial distribution is used to model the random behavior of data which results in one of two possible outcomes, typically labeled as success or failure. Another approach counts the number of "successes" per unit (that is, per interval of time, length, area, etc.). The counts per unit or counts per interval may be modeled by a Poisson distribution. Let Y be the random variable associated with such a count, and let λ be the appropriate expected rate of occurrences. The probability function for Y is

$$p(y) = \left\{ \begin{array}{cl} \frac{\lambda^y}{y!} \exp(-\lambda) & \text{for y = 0, 1, 2,... and } \lambda > 0 \\ 0 & \text{otherwise} \end{array} \right\}$$

If Y follows a Poisson distribution, then

$$\mu = E(Y) = \lambda,$$
$$\sigma^2 = \lambda, \text{ and}$$
$$\sigma = \sqrt{\lambda}.$$

Example 3.3 Nuclear Plant Pump Failures

Safety studies indicate that nuclear power plants depend heavily upon the reliability of their pumps. Let Y be the number of pump failures over a ten year period. The rate of pump failure over this period is believed to be 1.02 failures in this kind of application every ten years.

Follow these steps to determine the probability of at least one failure occurring over the ten year span.

Step 1. Enter Patterned Data.

Choose **Calc**>**Make Patterned Data**>**Simple Set of Numbers**. Type *Failures* in the Store patterned data in: text box. Place 0 in the From first value: text box and 10 in the To last value: text box. Choose **OK**.

Step 2. Construct the cumulative distribution.

Choose **Calc**>**Probability Distributions**>**Poisson**. Darken the Cumulative probability option button. Place 1.02 in the Mean: text box. Place Failures in the Input column: text box. Choose **OK**.

The Minitab Output

Cumulative Distribution Function

`Poisson with mu = 1.02000`

x	P(X <= x)
0.00	0.3606
1.00	0.7284
2.00	0.9160
3.00	0.9798
4.00	0.9960
5.00	0.9993
6.00	0.9999
7.00	1.0000
8.00	1.0000
9.00	1.0000
10.00	1.0000

Figure 3.7

The Minitab output, shown in Figure 3.7, indicates the resulting cumulative distribution function. The table indicates that the probability of 0 (or fewer) failures for pumps of this type in this kind of application is 0.3606. The probability of at least one failure is $1 - P(Y < 1) = 1 - P(Y = 0) = 1 - 0.3606 = .6394$. Consequently, at least failure is expected to occur over 63% of the time for this type of pump.

Plant management feels that they need to purchase a second pump to serve as backup if the chances of more than 2 failure is over 1%. The probability of more than 2 failures is $1 - P(Y \leq 2) = 1 - 0.9160 = .084$. As a result, plant management needs to purchase a backup pump.

The mean, variance, and standard deviation of the poisson probability distribution are: $\mu = \lambda$, $\sigma^2 = \lambda$, and $\sigma = \sqrt{\lambda}$. In this example,

$$\mu = \lambda = 1.02,$$
$$\sigma^2 = \lambda = 1.02, \text{ and}$$
$$\sigma = \sqrt{\lambda} = \sqrt{1.02} = 1.01.$$

Follow these steps to compare the theoretical values to the sampling distribution.

Step 1. Generate random data.

Choose **Calc**>**Random Data**>**Poisson**. Place 1000 in the Generate rows of data text box. Type *Data* in the Store in column(s): text box. Place 1.02 in the Mean: text box. Choose **OK**.

Step 2. Describe the data.

Choose **Stat**>**Basic Statistics**>**Descriptive Statistics**. Place Data in the Variables: text box. Choose **OK**.

The Minitab Output

Descriptive Statistics

Variable	N	Mean	Median	Tr Mean	StDev	SE Mean
Data	1000	1.0350	1.0000	0.9611	1.0014	0.0317

Variable	Min	Max	Q1	Q3
Data	0.0000	5.0000	0.0000	2.0000

Figure 3.8

The Minitab output, shown in Figure 3.8, indicates that the mean of the sample data is 1.0350 and the standard deviation is 1.0014. Compare these values for the sampling distribution to the theoretical mean and standard deviation for the Poisson probability distribution: 1.02 and 1.01, respectively. Your output may differ since this data is <u>random</u> data.

Exercises

3.1 Consider a process for making nickel battery plates that has an operator who successfully meets the weight specification only 20% of the time. Let Y be the number of times she successfully meets the specification on her next three attempts. If we assume that each attempt is independent of the other attempts, then the following table summarizes the probability function describing Y.

y_i	0	1	2	3
$p(y_i)$	0.512	0.384	0.096	0.008

Follow these steps to construct the cumulative distribution function of Y:

Step 1. Enter Data.

Enter the data under y_i into column C1. Enter the probabilities under p(y_i) into column C1. Name column C1 as y and column C2 as $p(y)$.

Step 2. Construct the cumulative distribution.

Choose **Calc**>**Probability Distributions**>**Discrete**. Darken the Cumulative Probability option button. Place y in the Values in: text box and P(y) in the Probabilities in: text box. Place y in the Input column: text box and CumProb in the Optional storage: text box. Choose **OK**.

Read the Minitab output to answer the following questions.
Find the probability that she is successful fewer than two times.
Find the probability that she is successful more than one time.

Follow these steps to determine the theoretical mean, variance and standard deviation for the number of times she is successful.

Step 1. Find the theoretical mean, using the equation $\sum y \cdot p(y)$.

Choose **Calc**>**Calculator**. Type *mu* in the Store result in variable: text box. Select Statistics from the Functions: drop down dialog box. Place Sum(y * 'P(y)') in the Expression: text box. Choose **OK**.

Step 2. Find the theoretical variance, using the equation $\sum y^2 \cdot p(y) - \mu^2$.

Choose **Calc**>**Calculator**. Type *Variance* in the Store result in variable: text box. Place SUM(y**2 * 'P(y)') - mu**2 in the Expression: text box. Choose **OK**.

Step 3. Find the theoretical standard deviation by taking the square root of the variance.

Choose **Calc**>**Calculator**. Type *StDev* in the Store result in variable: text box. Select Arithmetic from the Functions: drop down dialog box. Select Square root from the Functions: list box. Place SQRT(Variance) in the Expression: text box. Choose **OK**.

Follow these steps to compare the theoretical values to the sampling distribution.

Step 1. Generate random data.

Choose **Calc**>**Random Data**>**Discrete**. Place 1000 in the Generate rows of data text box. Type *Data* in the Store in column(s): text box. Place y in the Values in: text box. Place P(y) in the Probabilities in: text box. Choose **OK**.

Step 2. Describe the data.

Choose **Stat**>**Basic Statistics**>**Descriptive Statistics**. Place Data in the Variables: text box. Choose **OK**.

3.2 A pencil company has four extruders for making pencil lead. The maintenance manager has determined from historical data that the following table describes the distribution of the number of extruders down (out of operation) on any given day.

y_i	0	1	2	3	4
$p(y_i)$	0.5	0.3	0.1	0.05	0.05

a. Find the probability that three or more extruders are down.

b. Find the probability that less than one extruder is down.

c. Find the expected number of extruders down (theoretical mean).

d. Find the population variance and population standard deviation for the number of extruders down (theoretical variance and standard deviation).

3.3 An injection molding process for making detergent bottles uses four different machines. The following table describes the distribution for number of machines operating at any given time.

y_i	0	1	2	3	4
$p(y_i)$	0.005	0.010	0.035	0.050	0.900

a. Find the probability that two or less machines are running.

b. Find the probability that at least one machine is running.

c. Find the expected number of machines running (theoretical mean).

d. Find the theoretical variance and the standard deviation for the number of machines running.

3.4 A sales engineer for a manufacturer of high speed grinding equipment has just returned from visiting five possible clients. She believes that the following table describes the distribution for the number of sales she will make.

y_i	0	1	2	3	4	5
$p(y_i)$	0.05	0.30	0.30	0.20	0.10	0.05

a. Find the probability that she makes more than 3 sales.

b. Find the probability that she makes 2 or more sales.

c. Find the expected number of sales (theoretical mean).

d. Find the theoretical variance and the standard deviation for the number of sales.

3.5 A manufacturer of nickel-hydrogen batteries discovered a problem with "blisters" on its nickel plates. These blisters cause the resulting battery cell to short out prematurely. During a specific production period, 8.5% of the plates exhibited blisters within 50 test cycles. During this period, the company made a series of test cells, each using ten plates.

a. Find the probability that none of the ten plates blister.

b. Find the probability that exactly one blisters.

c. Find the expected number of plates that blister, the variance, and the standard deviation.

Follow these steps to answer questions a and b.

Step 1. Enter Patterned Data.

Choose **Calc**>**Make Patterned Data**>**Simple Set of Numbers**. Type *Blisters* in the Store patterned data in: text box. Place 0 in the From first value: text box and 10 in the To last value: text box. Choose **OK**.

Step 2. Calculate Probability Densities.

Choose **Calc**>**Probability Distributions**>**Binomial**. Darken the Probability option button. Place 10 in the Number of trials: text box. Place .085 in the Probability of success: text box. Place Blisters in the Input column: text box. Choose **OK**.

Read the Minitab output to answers questions a and b.

3.6 Atwood (1986) studied the failure of pumps used in standby safety systems for commercial nuclear-power plants. The number of pumps in a safety system ranged from 2 to 8 depending upon the specific nature of the system. He found that the probability that a randomly selected pump failed to run after starting was 0.16.

a. Typically, low-pressure coolant injection systems use four pumps. Consider a periodic inspection of this system which tests each pump.

i. Find the probability that all four fail a periodic inspection.

ii. Find the probability that at least one fails.

iii. Use simulation to compare the theoretical values to the sampling distribution. (Find the expected number of pumps which fail, the variance, and the standard deviation.)

b. Suppose a highly critical system uses 8 pumps. Consider a periodic inspection of this system which tests each pump.

 i. Find the probability that none of the pumps fail.

 ii. Find the probability that exactly two pumps fail.

 iii. Use simulation to compare the theoretical values to the sampling distribution. (Find the expected number of pumps which fail, the variance, and the standard deviation.)

3.7 Metal casting processes are notoriously slow and expensive. A common problem facing many older casting processes is "flashing." In casting, liquid metal is shot into a mold and rapidly cooled. A flash commonly forms on the piece at the spot in the mold where the metal flows. Sanding corrects minor problems with flashing. Severe problems may require scrapping the part. A particular casting operation historically has scrapped 10 flashing. Consider the next ten parts cast by this process, which represents about one hour of production.

 a. Find the probability that this company must scrap exactly two of these parts.

 b. Find the probability that it must scrap at least one part.

 c. Use simulation to compare the theoretical values to the sampling distribution. (Find the expected number of parts scrapped, the variance, and the standard deviation.)

3.8 An automobile manufacturer gives a 5-year/60,000 mile warranty on its drive train. Historically, 7% of this manufacturer's automobiles have required service under this warranty. Consider a randomly sample of 15 cars.

 a. Find the probability that exactly one car requires service under the warranty.

 b. Find the probability that more than one car requires service under the warranty.

 c. Find the expected time number of cars that require service, the variance and the standard deviation.

3.9 A civil engineering professor assigns a bridge building project using Popsicle sticks each semester. To get a passing grade, the structure must support at least 20 lb. Historically, 10% of the student bridges fail to support 20 lb. Assume that the current class of 15 students forms a random sample.

 a. Find the probability that everyone passes the project.

 b. Find the probability that at least one person fails the project.

 c. Find the expected number of students who pass, the variance and the standard deviation.

 d. Find the expected number of students who fail, the variance and the standard deviation.

3.10 If we reduce the times between industrial accident data to the number of accidents each month, then we can well model the data by a Poisson distribution with an accident rate of 1.5 per month.

 a. Find the probability that no accidents occur in a given month.

 b. Find the probability that at least one accident occurs in a given month.

c. Historically, DuPont management has placed the highest priority on safety and typically will re-assign any plant manager whose facility has an excessive number of accidents. Suppose that management begins to consider re-assignment once a facility has five accidents in a month. Find the probability that this facility has exactly five accidents in a given month.

Follow these steps to answer the above questions.

Step 1. Enter Patterned Data.
Choose **Calc**>**Make Patterned Data**>**Simple Set of Numbers**. Type *Accidents* in the Store patterned data in: text box. Place 0 in the From first value: text box and 10 in the To last value: text box. Choose **OK**.

Step 2. Calculate probability densities.
Choose **Calc**>**Probability Distributions**>**Poisson**. Darken the Probability option button. Place 1.5 in the Mean: text box. Place Accidents in the Input column: text box. Choose **OK**.

Step 3. Construct the cumulative distribution.
Choose **Calc**>**Probability Distributions**>**Poisson**. Darken the Cumulative probability option button. Place 1.5 in the Mean: text box. Place Accidents in the Input column: text box. Choose **OK**.

3.11 Nelson (1987) discusses a situation where a process historically has averaged 2.6 flaws per 1000 meter lengths of wire.
   a. Find the probability that a 1000-meter length of wire has one or less flaws.
   b. Find the probability that a 1000-meter length of wire has more than two flaws.
   c. Use simulation to compare the theoretical values to the sampling distribution. (Find the mean number of flaws per 1000-meter length of wire, the variance, and the standard deviation.)
   d. Consider a 500-meter length of wire.
      i. Find the probability that it has no flaws.
      ii. Use simulation to compare the theoretical values to the sampling distribution. (Find-the expected number of flaws, the variance, and the standard deviation.)

3.12 Kalbfleisch, Lawless, and Robinson (1991) modeled the number of warranty claims within one year of purchase for a particular system on a single car model with a Poisson distribution with a rate of 0.75 claims per vehicle. Consider a randomly selected automobile.
   a. Find the probability that this automobile has no claims within one year.
   b. Find the probability that this automobile has exactly three claims within one year.
   c. Use simulation to compare the theoretical values to the sampling distribution. (Find the expected number of claims, the variance, and the standard deviation.)

3.13 The manufacture of silicon wafers used in integrated circuits requires the removal of contaminating particles of a certain size. Yashcin (1995) studied a rinsing process for these wafers. This process rinses batches of 20 wafers with deionized water. The process then dries these wafers by spinning off the water droplets. Prior to loading the wafers in the rinser/dryer, production personnel count the number of contaminating particles. This count provides feedback on the cleanliness of the manufacturing environment. These counts are well modeled by a Poisson distribution with a rate of six particles per wafer. Consider a randomly selected wafer.

   a. Find the probability that this wafer has at least one particle.
   b. Find the probability that this wafer has exactly six particles.
   c. Use simulation to compare the theoretical values to the sampling distribution. (Find the expected number of particles, the variance, and the standard deviation.)

3.4   Continuous Random Variables

New Minitab Commands

1. **Calc>Probability Distributions>Exponential** Allows you to calculate probabilities, cumulative probabilities, and inverse cumulative probabilities for an exponential distribution. In this section, you will enter a mean in an appropriate text box, using this command to calculate cumulative probabilities.

2. **Graph>Plot** - Produces a scatter plot that shows the relationship between two variables. In this section you will produce scatter plots for a number of variables.

   a. **Annotation>Title** - Creates titles for a graph, placed at the top of the figure region, above the data region. You can create as many titles as you want, specifying different attributes for each title in the corresponding attributes rows (font, color, size, and so on). In this section, you will place a first title on a graph of a probability density function.

Continuous Random Variables

Let Y be a random variable. Y is said to be a continuous random variable if $F(y) = P(Y \leq y)$ is continuous for all values of y. For example, time is typically considered to be a continuous random variable.

Example 3.4 Times Between Industrial Accidents

Lucas (1985) studied the times between accidents for a ten year period at a DuPont facility. These times may be modeled by an *exponential distribution* with the following pdf:

$$f\{y\} = \left\{ \begin{array}{cc} \lambda \exp\left(-\lambda y\right) & y>0 \text{ and } \lambda > 0 \\ 0 & \text{otherwise} \end{array} \right\}$$

where λ is the accident rate (the expected number of accidents per day). The mean of an exponential distribution is given by $\frac{1}{\lambda}$. If $\lambda = 0.05$ accidents per day then the mean is $\frac{1}{\lambda} = \frac{1}{0.05} = 20$. Minitab requires that mean of 20 (days between accidents).

Follow these steps to construct the cumulative distribution function of Y:

Step 1. Enter Patterned Data.

Choose **Calc**>**Make Patterned Data**>**Simple Set of Numbers**. Type *y* in the **S**tore patterned data in: text box. Place 1 in the **F**rom first value: text box and 100 in the **T**o last value: text box. Choose **OK**.

Step 2. Construct the cumulative distribution.

Choose **Calc**>**Probability Distributions**>**Exponential**. Darken the **C**umulative probability option button. Place 20 in the **M**ean: text box. Place y in the Input co**l**umn: text box. Choose **OK**.

The Minitab Output

Cumulative Distribution Function

Exponential with mean = 20.0000

x	P(X <= x)
1.0000	0.0488
2.0000	0.0952
3.0000	0.1393
4.0000	0.1813
5.0000	0.2212
6.0000	0.2592
7.0000	0.2953
8.0000	0.3297
9.0000	0.3624
10.0000	0.3935
. . .	. . .
100.0000	0.9933

Figure 3.9

The Minitab output, as shown in Figure 3.9, indicates the resulting cumulative distribution function. The table indicates that the probability of ten or fewer days between accidents is 0.3935.

If you use the PgDn key or the scroll bar at the right of the Minitab Session Window to scroll down to 80 days, you can see that the $P(y \leq 80) = 0.9817$. Therefore, the probability that $P(y > 80) = 1 - P(y \leq 80) = 1 - 0.9817 = 0.0182$. As a result, the probability of going more than 80 days without an accident is fairly low.

The Relationship of p(y), f(y), and a Stem-and-Leaf
Formal statistical analyses require certain assumptions about the underlying distribution from which the data come. Appropriate data displays provide a quick and easy way to check on the assumption regarding the "shape" of the data distribution. For a discrete random variable, the probability function, p(y), and for a continuous random variable, the probability density function, f(y), define the theoretical shape for the stem and leaf display if the data come from the assumed distribution.

Example 3.4 - Continued - Times Between Industrial Accidents
   Follow these steps to graph the probability density function for this exponential distribution.
 Step 1. Calculate Probability Densities.
         Choose **Calc**>**Probability Distributions**>**Exponential**. Darken the Probability density option button. Place 20 in the Mean: text box. Place y in the Input column: text box. Place Probability in the Optional storage: text box. Choose **OK**.
 Step 2. Graph the probability density function.
         Choose **Graph**>**Plot**. Place Probability in the Y Graph variable(s): text box and y in the X Graph variable(s): text box.

         Enter a Title.
         Choose **Annotation**>**Title**. Place the title Theoretical Distribution for the Times Between Industrial Accidents in the first row of the Title text box. Choose **OK**.
         Choose **OK**.

The Minitab Output

Theoretical Distribution for the Times Between Industrial Accidents

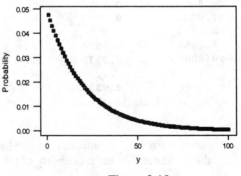

Figure 3.10

   To see if this distribution provides a reasonable model for the data we will look at a histogram (stem-and-leaf display) of the (simulated) data in a little while.

Expected Values of Continuous Random Variables

Minitab can be use to simulate probability distributions. Recall that the mean of the exponential distribution is given by $\frac{1}{\lambda}$ and the standard deviation is given by $\frac{1}{\lambda}$. In this example, with $\lambda = .05$, the mean is 20 and the standard deviation is $\frac{1}{\lambda} = 20$.

Follow these steps to simulate the exponential distribution:

Step 1. Generate random data.

> Choose **Calc**>**Random Data**>**Exponential**. Place 1000 in the Generate rows of data text box. Type *Exponential* in the Store in column(s): text box. Place 20 in the Mean: text box. Choose **OK**.

Step 2. Describe the data.

> Choose **Stat**>**Basic Statistics**>**Descriptive Statistics**. Place Exponential in the Variables: text box. Choose **OK**.

The Minitab Output

Descriptive Statistics

Variable	N	Mean	Median	Tr Mean	StDev	SE Mean
Exponent	1000	19.358	13.330	17.196	19.113	0.604

Variable	Min	Max	Q1	Q3
Exponent	0.002	136.035	5.793	27.244

Figure 3.11

The Minitab output, as shown in Figure 3.11, indicates that the mean of the sample data is 19.358 and the standard deviation is 19.113. Your output may differ since this data is <u>random</u> data.

Follow this sequence of steps to construct the histogram of the simulated data.

Choose **Graph**>**Histogram**. Place Exponential in the Graph variables: text box.

Enter a Title.
Choose **Annotation**>**Title**. Type *Simulated Data for the Times Between Industrial Accidents* in the first line of the Title text box. Choose **OK**.   .
Choose **OK**.

The Minitab Output

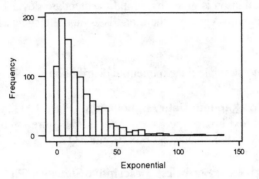

Simulated Data for the Times Between Industrial Accidents

Figure 3.12

This histogram, as shown in Figure 3.12, when compared to the graph of the probability density function, indicates that the exponential distribution provides a reasonable basis for modeling these times.

Some Important Continuous Distributions

New Minitab Commands

1. **Calc>Probability Distributions>Uniform.** Allows you to calculate probabilities, cumulative probabilities, and inverse cumulative probabilities for a uniform distribution. In this section, you will use this command to calculate cumulative probabilities in Exercise 3.17.

2. **Calc>Probability Distributions>Weibull.** Allows you to calculate probabilities, cumulative probabilities, and inverse cumulative probabilities for the Weibull distribution. In this section, you will use this command to calculate both probability densities and cumulative probabilities in Exercise 3.19.

Frequently engineers pay attention to three other important continuous probability distributions: the Uniform, Weibull and Gamma distributions. These distributions are often used to model times between events (such as failures) and interarrival times. Thus, these distributions play an important role in reliability and queuing studies.

Distribution	Density Function
Uniform	$\begin{cases} \dfrac{1}{b-a} & a \le y \le b \\ 0 & \text{otherwise} \end{cases}$
Weibull	$\begin{cases} \lambda\beta(\lambda y)^{\beta-1}\exp\left[-(\lambda y)^{\beta}\right] & y>0;\ \lambda>0;\ \beta>0 \\ 0 & \text{otherwise} \end{cases}$
Gamma	$\begin{cases} \dfrac{\lambda^{\alpha}y^{\alpha-1}}{\Gamma(\alpha)}\exp(-\lambda y) & y>0;\ \lambda>0;\ \beta>0 \\ 0 & \text{otherwise} \end{cases}$

The exponential distribution is a special case of both the gamma ($\alpha = 1$) and the Weibull distribution ($\beta = 1$).

Exercises

3.14 Davis and Lawrance (1989) present time to failure data for 171 automobile tires. Baltazar-Aban and Pena (1995) modeled these data with an exponential distribution with a failure rate of .004 tires per hour.
  a. Find the probability that a randomly selected tire fails in the first 100 hours of testing.
  b. Find the probability that a randomly selected tire fails between 50 and 150 hours of testing.
  c. Find the probability that a randomly selected tire survives more than 200 hours.
  d. Find the expected time until failure for a randomly selected tire.
  e. Find the theoretical variance and the standard deviation for the time to failure for a randomly selected tire.

Follow these steps to answer questions a, b and c.
  Step 1. Enter Patterned Data.
      Choose **Calc**>**Make Patterned Data**>**Simple Set of Numbers**. Type *Hours* in the Store result in column: text box. Place 1 in the From first value: text box and 500 in the To last value: text box. Choose **OK**.
  Step 2. Construct the cumulative distribution.
      Choose **Calc**>**Probability Distributions**>**Exponential**. Darken the Cumulative Probability option button. Place the mean of the exponential distribution, $\frac{1}{\lambda} = \frac{1}{.004} = 250.0$ in the Mean: text box. Darken the option button for Input column:. Place Hours in the Input column: text box. Choose **OK**.
      Read the Minitab output from the Session window to answer questions a, b and c.

Follow these steps to compare the compare the theoretical values to the sampling distribution of the time until failure for a randomly selected tire.
  Step 1. Generate random data.
      Choose **Calc**>**Random Data**>**Exponential**. Place 1000 in the Generate rows of data text box. Type *Exponential* in the Store in column(s): text box. Place 250 in the Mean: text box. Choose **OK**.
  Step 2. Describe the data.
      Choose **Stat**>**Basic Statistics**>**Descriptive Statistics**. Place Exponential in the Variables: text box. Choose **OK**.
      Read the Minitab output from the Session window to compare the

theoretical values to the sampling distribution.

3.15 Miyamura (1982) modeled the life times for the electromagnetic valve used for starting the idle-up actuator of an air conditioner by an exponential distribution with a rate of .05 failures per million revolutions. Consider a randomly selected valve.

   a. Find the probability that this valve fails within the first half million revolutions.

   b. Find the probability that this valve lasts longer than 3 million revolutions.

   c. Find the expected time until failure for this valve.

   d. Find the variance and the standard deviation for the time until failure for this valve.

3.16 The maintenance manager at a chemical facility knows that the times between repairs, Y, for a specific chemical reactor are well modeled by the following distribution.

$$f(y) = .01e^{-.01y}$$

for y > 0 and 0 otherwise. (Hint: Identify λ)

   a. Find the probability that Y is less than 30.

   b. Find the probability that Y is greater than 15.

   c. Find the probability that Y is exactly equal to 100.

   d. Find the probability that Y is between 50 and 150.

3.17 Rather than collecting the actual times between accidents, Lucas could have reduced the data to the number of accidents per month. Suppose that we only know that one accident occurred in a given month and that we can well model the actual times between accidents by an exponential distribution. In this case, given that we know an accident occurred at some time during the month, the time within the month it happened follows a uniform distribution. In general, the pdf for a uniform distribution is

$$f(y) = \left\{ \begin{array}{cc} \frac{1}{b-a} & a \le y \le b \\ 0 & \text{otherwise} \end{array} \right\}$$

This pdf essentially says that all the values within this interval are equally likely to occur.

   a. Find the cdf for this distribution.

   b. For the time between accidents data, assume that a month has 30 days and that an accident occurs in that month.

      i. Find the probability that the accident occurred on or before the 15th.

      ii. Find the probability it occurred after the 25th.

Follow these steps to find the cdf for this distribution.

Step 1. Enter Patterned Data.

Choose **Calc**>**Make Patterned Data**>**Simple Set of Numbers**. Type *Days* in the **S**tore result in column: text box. Place 1 in the **F**rom first value: text box and 30 in the **T**o last value: text box. Choose **OK**.

Step 2. Construct the cumulative distribution.

Choose **Calc**>**Probability Distributions**>**Uniform**. Darken the **C**umulative Probability option button. Place 1 in the Lo**w**er endpoint: text box. Place 30 in the **U**pper endpoint: text box. Darken the option button for Input column:. Place Days in the Input co**l**umn: text box. Choose **OK**.

Read the Minitab output from the Session window to answer questions above.

3.18 Engineers often use the uniform distribution to model the arrival time of some event given that the event did occur within some interval. For example, production knows that a particular pump failed some time between 1:00 and 3:00 p.m. Given that we know it failed at some time during this period, the pdf for the specific time within the period is

$$f(y) = \left\{ \begin{array}{cc} \frac{1}{b-a} & a \leq y \leq b \\ 0 & \text{otherwise} \end{array} \right\}$$

where a = 1 and b = 3. This pdf essentially says that all the times within this interval are equally likely to occur.

a. Find the mean for this distribution.

b. Find the variance and standard deviation for this distribution.

c. Find the probability that the pump failed after 1:30 p.m.

3.19 The Weibull distribution is probably the most widely used distribution in reliability. The pdf for the Weibull is given by

$$f(y) = \left\{ \begin{array}{cc} \lambda\beta(\lambda y)^{\beta-1} \exp\left[-(\lambda y)^{\beta}\right] & y > 0; \lambda > 0; \beta > 0 \\ 0 & \text{otherwise} \end{array} \right\}$$

For $\beta = 1.0$, the Weibull simplifies to the exponential distribution. Padgett and Spurrier (1990) us a Weibull with $\lambda = 0.4$ and $\beta = 2$ to model the breaking strengths of carbon fibers used in fibrous composite materials. These fibers measure 50 mm in length and 7-8 microns in diameter. Periodically, the manufacturer selects random samples of five fibers and tests their breaking stresses.

The first question:

a. Specification suggest that 99% of the fibers must have a breaking strength of at least 1.2 GPa (giga-Pascals). Find the probability that the breaking strength is less than 1.2 GPa.

Follow these steps to answer the above question.

Step 1. Enter Patterned Data.

Choose **Calc**>**Make Patterned Data**>**Simple Set of Numbers**. Type *GPa* in the Store result in column: text box. Place 0 in the From first value: text box and 6 in the To last value: text box. Place 0.1 in the Increment: text box. Choose **OK**.

Step 2. Construct the probability densities.

Choose **Calc**>**Probability Distributions**>**Weibull**. Darken the option button for Probability density. Place 2 in the Shape parameter: text box. Place $\frac{1}{\lambda} = \frac{1}{0.4} = 2.5$ in the Scale parameter: text box. Darken the option button for Input column:. Place GPa in the Input column: text box.. Type *pdf* in the Optional storage: text box. Choose **OK**. (The Weibull density function contains two parameters, λ and β. The scale parameter, λ, reflects the size of the units in which the random variable y is measured. The parameter β is the shape parameter, generating the curve to model the distribution.)

Step 3. Construct the cumulative distribution.

Choose **Calc**>**Probability Distributions**>**Weibull**. Darken the Cumulative Probability option button. Place 2 in the Shape parameter: text box. Place $\frac{1}{\lambda} = \frac{1}{0.4} = 2.5$ in the Scale parameter: text box. Darken the option button for Input column:. Place GPa in the Input column: text box.. Place cdf in the Optional storage: text box. Choose **OK**.

Read the Minitab output from the Session window to answer the above question.

The second question:

b. The breaking stresses in GPa for the samples follow. Compare the plot of the pdf with a stem-and-leaf display of the actual strengths and comment.

1.4	3.7	3.0	1.4	1.0	2.8	4.9	3.7	1.8	1.6
3.2	1.6	0.8	5.6	1.7	1.6	2.0	1.2	1.1	1.7
2.2	1.2	5.1	2.5	1.2	3.5	2.2	1.7	1.3	4.4
1.8	0.4	3.7	2.5	0.9	1.6	2.8	4.7	2.0	1.8
1.6	1.1	2.0	1.6	2.1	1.9	2.9	2.8	2.1	3.7

Follow these steps to answer the above question.

Step 1. Enter data.

Go to the data window. Entering the data in column C4. Name column C4 as *Stresses*.

Step 2. Plot the pdf.

Choose **Graph**>**Plot**. Place pdf in the Y Graph variables: text box. Place GPa in the X Graph variables: text box. Choose **OK**.

Step 3. Construct the stem-and-leaf display.

Choose **Graph**>**Character Graphs**>**Stem-and-Leaf.** Place GPa in the Variables: text box. Place .1 in the Increment: text box. Choose **OK**.

Compare the plot of the pdf with the stem-and-leaf display of the actual strengths and comment on the comparison.

3.20 All batteries have definite "shelf" lives, i.e. batteries begin to deteriorate in storage. Morris (1987) used a Weibull distribution (see the previous exercise) with $\beta = 2$ and $\lambda = 0.1$ to model the storage time required for long-life Li/SO_4 batteries to become acceptable for use.

a. Find the probability that a randomly selected battery becomes unacceptable between 12 and 18 months.

3.5   The Normal Distribution

New Minitab Commands

1. **Calc**>**Probability Distributions**>**Discrete** Allows you to calculate probabilities, cumulative probabilities, and inverse cumulative probabilities for a discrete distribution. In this section, you will enter a discrete probability distribution in the Data window, using this command to calculate cumulative probabilities.

2. **Calc**>**Probability Distributions**> **Normal** - Allows you to calculate the probability densities, cumulative probabilities, and inverse cumulative probabilities for a normal distribution. For the continuous distributions, such as the normal distribution, Minitab calculates the continuous probability density function.

a. **Cumulative probability:** In this section, you will darken the Cumulative probability: option button in order to determine the area under the normal probability density function.

Among all of the distributions used in classical statistics, the single most important is the normal distribution, which is often described as the "bell shaped" distribution or curve. Areas under the standard normal curve can be found with the aid of standard normal tables found in most textbooks. Minitab can also be used to find areas under the standard normal curve. Minitab provides the area that lies to the *left* of a specified value of z.

Example 3.5 - Production at a Titanium Dioxide Facility

A major titanium dioxide (white pigment) facility uses two production lines. Historically, the total daily production of both lines together is approximately normally distributed with a mean of 500 tons and a standard deviation of 50 tons. Find the probability that the production exceeds 600 tons on any one randomly selected day.

Follow this step to find the probability that production exceeds 600 tons on any one randomly selected day.

Choose **Calc**>**Probability Distributions**>**Normal**. Darken the **C**umulative probability option button. Place 500.0 in the **M**ean: text box. Place 50.0 in the **S**tandard deviation: text box. Darken the option button for Input constant. Place 600 in the Input constant: text box. Choose **OK**.

The Minitab Output

Cumulative Distribution Function

```
Normal with mean = 500 and standard deviation = 50


          x        P( X <= x)
    600.0000          0.9772
```
Figure 3.13

The Minitab output, shown in Figure 3.13, indicates that the probability of production less than 600 tons is 0.9772. The probability of production exceeding 600 tons on any one randomly selected day is 1 - 0.9772 = 0.228. As a result the facility should exceed 600 tons only about 2% of the time.

The production superintendent believes that "typical" performance is between 480 and 520 tons.

Follow these steps to find the probability that production falls with the interval from 480 to 520 tons.

Step 1. Enter data.

Go to the data window and enter 480 and 520 in rows one and two of column C1.

Step 2. Construct the cumulative distribution.

Choose **Calc**>**Probability Distributions**>**Normal**. Darken the **C**umulative probability option button. Place 500.0 in the **M**ean: text box. Place 50.0 in the **S**tandard deviation: text box. Darken the option button for Input column. Place C1 in the Input column: text box. Choose **OK**.

The Minitab Output

Cumulative Distribution Function

Normal with mean = 500 and standard deviation = 50

x	P(X <= x)
480.0000	0.3446
520.0000	0.6554

Figure 3.14

The Minitab output, as shown in Figure 3.14, indicates that the probability of production less than 480 tons is 0.3446 and less than 520 tons is 0.6554. The probability that production falls within the interval from 480 to 520 tons on any one randomly selected day is $0.6554 - 0.3446 = .3108$.

Exercises

3.21 Wasserman and Wadsworth (1989) discuss a process for the manufacture of steel bolts which continuously feed an assembly line down stream. Historically, the thicknesses of these bolts follow a normal distribution with a mean of 10.0 mm and a standard deviation of 1.6 mm. Process supervision becomes concerned about the process if the thicknesses begin to get larger than 10.8 mm or smaller than 9.2 mm. Assume that the current process mean is 10.0 mm, and consider a randomly selected bolt.

　　a. Find the probability that the thickness of this bolt is between 9.2 mm and 10.8 mm.

　　b. Find the probability that the thickness of this bolt is less than 9.2 mm.

Follow these steps to answer the above questions.

　　Step 1. Enter Patterned Data.

　　　　　Choose **Calc**>**Make Patterned Data**>**Simple Set of Numbers**. Type *Thickness* in the Store result in column: text box. Place 9 in the From first value: text box and 11 in the To last value: text box. Choose **OK**.

　　Step 2. Construct the cumulative distribution.

　　　　　Choose **Calc**>**Probability Distributions**>**Normal**. Darken the Cumulative probability option button. Place 10.0 in the Mean: text box. Place 1.6 in the Standard deviation: text box. Place Thickness in the Input column: text box. Choose **OK**.

　　　　　Use the information in the Session window to answer questions one and two.

3.22 Canning (1993) studied the performance of a new photolithography process. This process deposits layers of material on silicon wafers. The thicknesses of one of the layers deposited followed a normal distribution with a standard deviation of .057. Since all thicknesses were measured relative to the nominal, the mean was 0. The lower and upper specifications were -0.15 and 0.15, respectively.

   a. Find the probability that any given layer deposited by the new process falls within the specifications.

   b. Find the value for the mean thickness required to make the probability of exceeding the upper specification limit be less than 1%.

3.23 Runger and Pignatiello (1991) consider a plastic injection molding process for a part with a critical width dimension which follows a normal distribution with a historic mean of 100 and historic standard deviation of 8. They monitor this process with a control chart with limits of 90 and 110. Assume that the mean width is 100, and consider a randomly selected part.

   a. Find the probability that the width of this part is larger than 110.

   b. Find the probability that the width is smaller than 90.

   c. Runger and Pignatiello also use "warning" limits of 99 and 101 to help monitor this process. Find the probability that the width is between 99 and 101.

3.24 An ethanol-water distillation column historically produces yield which are well modeled by a normal distribution with a mean of 70 volume per cent and a standard deviation of 2.

   a. Find the probability that a yield exceeds 75%.

   b. Find the probability that a yield is between 67% and 73%.

3.25 The volumes delivered by a nominal 20 oz soft drink bottling process follow a normal distribution with a mean of 20.2 oz and a standard deviation of .07.

   a. Find the probability that this process underfills a bottle.

   b. The bottle will overflow if the volume delivered exceeds 20.35 oz. Find the probability of an overflow.

3.26 The daily production of a sulfuric acid process is known to follow a normal distribution with a mean of 400 tons per day and a variance of 225 tons per day.

   a. Find the probability that today's production will be between 375 and 425 tons.

   b. Find the probability that today's production will be less than 360 tons.

3.6   Random Behavior of Means

New Minitab Commands

1. **Calc>Row Statistics** - Computes one value for each row in a set of columns. The statistic is calculated across the rows of the column(s) specified and the answers are stored in the corresponding rows of a new column. In this section, you will use this command to calculate row means.

Engineering methods require the collection of data and the capability of modeling the random behavior of the statistics obtained from the sample data. Minitab may be used to simulate the probability distribution of a statistic via the sampling distribution of the statistic.

Example 3.6 - Thicknesses of Silicon Wafers

Hurwitz and Spagon (1993) analyzed the performance of a planarization device which polishes silicon wafers to a high degree of smoothness. Historically, the thicknesses at the dead center of the wafer are approximately normally distributed with a mean of 3200 Angstroms with a standard deviation of 80 Angstroms. A random sample of size 23 wafers is selected. The sample mean of those 23 wafers is found to be 3232 Angstroms. Production management would like to know if there is evidence to suggest that this lot is thicker than normal.

Follow these steps to simulate the thicknesses of the wafers and provide some insight as to the sampling distribution of the sample mean.

Step 1. Generate random data.

Choose **Calc**>**Random Data**>**Normal**. Place 1000 in the Generate rows of data text box. Type *C1-C23* in the Store in column(s): text box. Place 3200 in the Mean: text box. Place 80 in the Standard deviation: text box. Choose **OK**.

Step 2. Calculate the means for each sample (row).

Choose **Calc**>**Row Statistics.** Darken the option button for Mean. Place C1-C23 in the Input variables: text box. Place Means in the Store result in: text box. Choose **OK**

Step 3. Plot a histogram of the sample means.

Choose **Graph**>**Histogram**. Place Means in the Graph variables: text box. Choose **OK**.

The Minitab Output

A Simulation of the Sampling Distribution of Sample Means

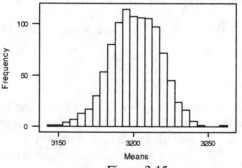

Figure 3.15

Step 4. Describe the sample means.

Choose **Stat**>**Basic Statistics**>**Descriptive Statistics**. Place Means in the Variables: text box. Choose **OK**.

The Minitab Output

Descriptive Statistics

Variable	N	Mean	Median	Tr Mean	StDev	SE Mean
Means	1000	3201.1	3201.4	3201.3	16.4	0.5

Variable	Min	Max	Q1	Q3
Means	3147.2	3259.6	3190.0	3213.2

Figure 3.16

Recall that the Central Limit Theorem states that if random samples of n observations are taken from an underlying population with mean μ and standard deviation σ, then the sampling distribution of sample means will be approximately normally distributed with a mean μ and standard deviation $\frac{\sigma}{\sqrt{n}}$. The Minitab output, as shown in Figure 3.15 and Figure 3.16, indicates that the mean of the distribution of sample means for n=23 is 3201.1. This is fairly close to the mean of the original distribution, 3200. The Minitab output shows that the standard deviation of the distribution of sample means for n=23 is 16.4. The Central Limit Theorem would have predicted $\frac{\sigma}{\sqrt{n}} = \frac{80}{\sqrt{23}} = 16.681$.

Follow this step to find the probability that the sample mean of 3232 Angstroms would suggest that this lot is thicker than normal.

Choose **Calc**>**Probability Distributions**>**Normal**. Darken the option button for Cumulative probability. Place 3200.0 in the Mean: text box. Place 16.681 in the Standard deviation: text box. Darken the option button for Input constant. Place 3232 in the Input constant: text box. Choose **OK**.

The Minitab Output

Cumulative Distribution Function

Normal with mean = 3200 and s.d. = 16.6810

x	P(X <= x)
3.23E+03	0.9725

Figure 3.17

The Minitab output, as shown in Figure 3.17, indicates the probability of a sample mean of 3232 or less is 0.9725. The probability of a sample mean exceeding 3232 is 1 - 0.9725 = 0.275. As a result the chances of observing a sample mean of 3232 or larger is less than 3%. There is some evidence to suggest that this particular lot is thicker than normal.

Normal Probability Plots

New Minitab Commands

1. <u>C</u>alc><u>Ca</u>lculator - Does arithmetic using an algebraic expression.
   a. Normal scores (NSCOR) - Calculates normal scores, which can be used to produce normal probability plots. In this section, you will use this command in constructing a normal probability plot. You can produce normal probability plots directly by using <u>Stat</u>><u>B</u>asic Statistics><u>N</u>ormality Test. You will also use this command in this section to produce normal probability plots.

2. <u>Stat</u>><u>B</u>asic Statistics><u>D</u>escriptive Statistics> <u>N</u>ormality Test - Generates a normal probability plot. The grid on the graph resembles the grids found on normal probability paper. The vertical axis has a probability scale; the horizontal axis, a data scale. A least-squares line is fit to the plotted points and drawn on the plot for reference. The line forms an estimate of the cumulative distribution function for the population from which data are drawn. Numerical estimates of the population parameters, μ and σ, are displayed with the plot. In this section, you will use this command to construct a normal probability plot to examine the assumption that the distribution from which these observations were drawn is normally distributed.

Normal probability plots can help us to determine whether we can reasonably model our data by a normal distribution. A substantial linear pattern in a normal probability plot suggests that the data follows a normal distribution. On the other hand, a systematic departure from a straight-line patter (such as curvature in the plot) casts doubt on the legitimacy of assuming a normal population distribution.

Within Minitab, you may construct normal probability plots. We will look at two techniques for constructing normal probability plots. Example 3.7 will use normal scores (NSCOR) to produce a normal probability plot, while Example 3.8 will produce a normal probability plot directly by using the <u>Stat</u>><u>B</u>asic Statistics><u>N</u>ormality Test command.

Example 3.7 - Thicknesses of Silicon Wafers - Continued

Hurwitz and Spagon (1993) analyzed the performance of a planarization device which polishes silicon wafers to a high degree of smoothness. The thicknesses for 23 wafers from a single lot, all measured at the center of the wafer were recorded:

3240	3200	3220	3210	3250	3220
3190	3190	3150	3160	3270	3180
3200	3270	3180	3300	3250	3330
3300	3280	3270	3270	3200	

Follow these steps to construct a normal probability plot to examine the assumption that the assumption that the distribution of the thicknesses of the silicon wafers from which these observations were drawn is normal. In this example, you will use normal scores to construct this normal probability plot.

Step 1. Enter data.

Enter the observations in column C1 of the Data window. Name column C1 as *Thickness*.

Step 2. Determine normal scores.

Choose **Calc**>**Calculator**. Type *NormalScores* in the Store result in variable: text box. Select Statistics from the Functions: drop down dialog box. Choose Normal scores from the Statistics functions list box, placing NSCOR(number) in the Expression: text box. Replace number in the NSCOR(number) in the expression with Thickness. Choose **OK**.

Step 3. Create the normal probability plot.

Choose **Graph**>**Plot**. Place NormalScores in the Y Graph variables: text box. Place Thickness in the X Graph variables: text box. Choose **OK**.

The Minitab Output

A Normal Probability Plot of Silicon Wafer Thicknesses

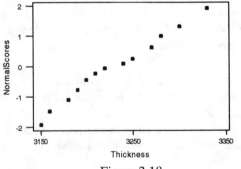

Figure 3.18

The Minitab output, as shown in Figure 3.18, indicates the data appear roughly to follow a straight line. We thus feel reasonably comfortable that the data come from a well-behaved distribution. Especially with a sample size of 23, we feel comfortable that the sample mean follows a normal distribution by the Central Limit Theorem.

Example 3.8 - Times Between Industrial Accidents - Continued

Lucas (1985) studied the times between accidents for a ten year period at a DuPont facility. DuPont historically has strongly emphasized the importance of safety in its operations and has always striven to reduce the number of accidents over time. During this period, 178 accidents occurred.. Table 2.5 in the text gives the data, which are the number of days since the previous accident.

Follow these steps to construct a normal probability plot to examine the assumption that the assumption that the distribution of the times between industrial accidents from which these observations were drawn is normal. In this example, you will use the **Stat**>**Basic Statistics**>**Normality Test** command to construct this normal probability plot.

Step 1. Enter data.

Enter the observations in column C1 of the Data window. Name column C1 as *Times*.

Step 2. Create the normal probability plot.

Choose **Stat**>**Basic Statistics**>**Normality Test**. Place Times in the **V**ariable: text box. Place Times Between Industrial Accidents in the **T**itle: text box. Choose **OK**.

The Minitab Output

Times Between Industrial Accidents

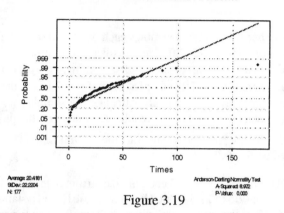

Figure 3.19

The Minitab output, as shown in Figure 3.19, indicates the data look nothing like a normal distribution. Earlier we noted that the exponential distribution provides a reasonable model for these data. As a result, we expect this plot to look quite different from a straight line.

Computer Exercises

3.27 Use a statistical software package to illustrate the Central Limit Theorem when sampling from a normal distribution.

a. Generate 1000 random samples each of size 4 from a normal distribution with a mean of 10 and a standard deviation of 10. Have the software calculate the sample mean for each sample. Plot a histogram of the sample means. Comment on the plot.

Follow these steps to perform the operation above.

Step 1. Name columns.

Place the pointer directly above row 1 in column 1 and name column 1 as *Data1*. Repeat the process, naming column 2 as *Data2*, column 3 as *Data3*, and column 4 as *Data4*.

Step 2. Generate random data.

Choose **Calc**>**Random Data**>**Normal**. Place 1000 in the Generate rows of data text box. Place Data1-Data4 in the Store in column(s): text box. Place 10 in the Mean: text box. Place 10 in the Standard deviation: text box. Choose **OK**.

Step 3. Calculate the means for each sample (row).

Choose **Calc**>**Row Statistics**. Darken the option button for Mean. Place Data1-Data4 in the Input variables: text box. Place Means in the Store result in: text box. Choose **OK**.

Step 4. Plot a histogram of the sample means.

Choose **Graph**>**Histogram**. Place Means in the Graph variables: text box. Choose **OK**.

b. Generate another 1000 random samples each of size 10 from the same normal distribution. Have the software calculate the sample mean for each sample and plot a histogram of the results. Comment on your plot.

c. Generate another 1000 random samples each of size 30 from the same normal distribution. Have the software calculate the sample mean for each sample and plot a histogram of the results. Comment on the plot.

3.28 Use a statistical software package to illustrate the Central Limit Theorem when sampling from a uniform distribution.

a. Generate 1000 random samples each of size 4 from a uniform distribution over the interval -7.5 to 27.5, which has a mean of 10 and a standard deviation of 10.1. Have the software calculate the sample mean for each sample and plot a histogram of the results. Comment on the plot.

b. Generate another 1000 random samples each of size 10 from the same uniform distribution. Have the software calculate the sample mean for each sample and plot a histogram of the results. Comment on the plot.

c. Generate another 1000 random samples each of size 30 from the same uniform distribution. Have the software calculate the sample mean for each sample and plot a histogram of the results. Comment on the plot.

3.29 Use a statistical software package to illustrate the Central Limit Theorem when sampling from an exponential distribution.

a. Generate 1000 random samples each of size 4 from an exponential distribution with $\lambda = 0.1$, which has a mean and a standard deviation of 10. Have the software calculate the sample mean for each sample and plot a histogram of the results. Comment on the plot.

b. Generate another 1000 random samples each of size 10 from the same exponential distribution. Have the software calculate the sample mean for each sample and plot a histogram of the results. Comment on the plot.

c. Generate another 1000 random samples each of size 30 from the same exponential distribution. Have the software calculate the sample mean for each sample and plot a histogram of the results. Comment on the plot.

Exercises

3.30 Kane (1986) discusses the concentricity of an engine oil seal groove. Concentricity measures the cross-sectional coaxial relationship of two cylindrical features. In this case, he studied the concentricity of an oil seal groove and a base cylinder in the interior of the groove. He measures the concentricity as a positive deviation using a dial indicator gauge. Historically, this process has produced an average concentricity of 5.6 with a standard deviation of 0.7. To monitor this process, he periodically takes a random sample of three measurements. If the average is greater than 6.8 or less than 4.3, he concludes that the process mean has shifted. Assume that the process mean is 5.6.

    a. Find the probability that on the next sample he concludes the process mean has shifted purely due to random chance.

    b. What did you assume in order to find this probability?

3.31 Runger and Pignatiello (1991) consider a plastic injection molding process for a part with a critical width dimension with a historic mean of 100 and historic standard deviation of 8. Periodically, clogs form in one of the feeder lines causing the mean width to change. As a result, the operator periodically takes random samples of size four. If the sample mean width of these four parts is either larger than 101.0 or smaller than 99.0, then he must immediately take another sample. Consider the next sample taken. Assume that the actual mean width is 100.

    a. Find the probability that the operator must take another sample immediately.

    b. What did you assume in order to find this probability?

3.7   The t-Distribution

New Minitab Commands

1. **Stat>Basic Statistics>1-Sample t** - Performs a one sample t-test or t-confidence interval for the mean. In this section, you will use this command to perform one sample t-tests.

The Central Limit Theorem states that if the sample size is sufficiently large then

$$Z = \frac{\overline{y}-\mu}{\frac{\sigma}{\sqrt{n}}}$$

follows a standard normal distribution. On the other hand if s is substituted for σ, then

$$\frac{\overline{y}-\mu}{\frac{s}{\sqrt{n}}}$$

follows a t distribution with n-1 degrees of freedom.

Example 3.9 - Packaged Weights

King (1992) discusses the net weights of a nominally 16 oz. packaged product. An inspector examined 20 such packages and found the sample mean weight to be 16.52 oz. and the sample standard deviation to be 0.16 ($\overline{x} = 16.52$, $s = 0.16$). **Follow these steps** to illustrate the distribution of the t-statistic and to find the probability of a sample mean of 16.52 oz. if the true process mean is 16.0 oz.

Step 1. Generate random data.

Choose **Calc**>**Random Data**>**Normal**. Place 1000 in the Generate rows of data text box. Type *C1-C20* in the Store in column(s): text box. Place 16.52 in the Mean: text box. Place 0.16 in the Standard deviation: text box. Choose **OK**.

Step 2. Calculate the means for each sample (row).

Choose **Calc**>**Row Statistics**. Darken the option button for Mean. Place C1-C20 in the Input variables: text box. Place Means in the Store result in: text box. Choose **OK**.

Step 3. Calculate the t-statistic for each sample mean.

Choose **Calc**>**Calculator**. Type *tstat* in the Store result in variable: text box. Place ('Means'-16)/(0.16/Sqrt(20)) in the Expression: text box. Choose **OK**.

Use {Ctrl}D to go to the data window. Look in the column named Means, locating a mean close to 16.52. Observe the tstat (t-statistic). A sample mean of 16.52 will have a t-statistic of approximately 14.53. The chances of seeing such an extreme value due to random chance are extremely remote. The evidence indicates that the true process mean is something larger than 16.0 oz.

Step 4. Plot a histogram of the t-statistic.

Choose **Graph**>**Histogram**. Place tstat in the Graph variables: text box. Choose **OK**.

The Minitab Output

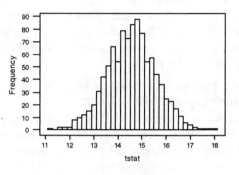

Figure 3.20

The Minitab output, as shown in Figure 3.20, indicates the sampling distribution of the sample t-statistics.

Computer Exercises

3.31 Use a statistical software package to illustrate the distribution of the t-statistic when sampling from a normal distribution.

    a. Generate 1000 random samples each of size 4 from a normal distribution with a mean of 10 and a standard deviation of 10. Have the software calculate the sample mean, the sample standard deviation, and the t-statistic for each sample. Plot a histogram of the t-statistics and comment on the plot.

    b. Generate another 1000 random samples each of size 10 from the same normal distribution. Have the software calculate the sample mean, the sample standard deviation, and the t-statistic for each sample. Plot a histogram of the results and comment on your plot.

    c. Generate another 1000 random samples each of size 30 from the same normal distribution. Have the software calculate the sample mean, the sample standard deviation, and the t-statistic for each sample. Plot a histogram of the results and comment on the plot.

3.32 Use a statistical software package to illustrate the distribution of the t-statistic when sampling from a uniform distribution.

    a. Generate 1000 random samples each of size 4 from a uniform distribution over the interval -7.5 to 27.5, which has a population mean of 10 and a population standard deviation of 10.1. Have the software calculate the sample mean, the sample standard deviation, and the t-statistic for each sample. Plot a histogram of the results and comment on the plot.

b. Generate another 1000 random samples each of size 10 from the same uniform distribution. Have the software calculate the sample mean, the sample standard deviation, and the t-statistic for each sample. Plot a histogram of the results and comment on the plot.

c. Generate another 1000 random samples each of size 30 from the same uniform distribution. Have the software calculate the sample mean, the sample standard deviation, and the t-statistic for each sample. Plot a histogram of the results and comment on the plot.

3.33 Use a statistical software package to illustrate the distribution of the t-statistic when sampling from an exponential distribution.

a. Generate 1000 random samples each of size 4 from an exponential distribution with $\lambda = 0.1$, which has a population mean and a population standard deviation of 10. Have the software calculate the sample mean, the sample standard deviation, and the t-statistic for each sample. Plot a histogram of the results and comment on the plot.

b. Generate another 1000 random samples each of size 10 from the same exponential distribution. Have the software calculate the sample mean, sample standard deviation, and the t-statistic for each sample. Plot a histogram of the results and comment on the plot.

c. Generate another 1000 random samples each of size 30 from the same exponential distribution. Have the software calculate the sample mean, the sample standard deviation, and the t-statistic for each sample. Plot a histogram of the results and comment on the plot.

Exercises

3.34 (EXO348) Pignatiello and Ramberg (1985) studied the heat treatment of leaf springs. In this process, a conveyor system transports leaf spring assemblies through a high temperature furnace. After this heat treatment, a high pressure press induces the curvature. After leaving the press, an oil quench cools the spring to near ambient temperature. An important quality characteristic of this process is the resulting free height of the spring, which has a target value of exactly eight inches. The following table summarizes the resulting leaf spring heights for a heating time of 23 seconds. Assume these data form a random sample.

7.5 7.6 7.5 7.5 7.6 7.5
7.6 7.6 7.8 7.6 7.8 7.6
7.6 7.6 7.4 7.2 7.2 7.3
7.6 7.8 7.7 7.8 7.5 7.6

a. Find the sample mean, sample variance, and sample standard deviation for these data.

Follow these steps to find the sample statistics.

Step 1. Enter data.
Enter the data in column 1. Name column 1 as *Heights*.

Step 2. Obtain descriptive statistics.
Choose **Stat**>**Basic Statistics**>**Descriptive Statistics**. Place Heights in the Variables: text box. Choose **OK.**

   b. Assume that the true population mean is 8 in. Find the observed value of the t-statistic. Use the rough rules of thumb to interpret your result.
   Follow this step to calculate the t-statistic.
   Choose **Stat**>**Basic Statistics**>**1-Sample t.** Place Heights in the Variables: text box. Darken the **T**est mean: option button. Place 8 in the **T**est mean: text box. Select the default value of not equal from the Alternative drop down dialog box. Choose **OK.**

   c. What did you assume in order to do the above informal analysis? Can you evaluate how well these data meet these assumptions? If yes, determine how comfortable you are with them. If not, explain why.
   Follow this step to construct a normal probability plot to examine the assumption that the assumption that the distribution of the leaf spring heights from which these observations were drawn is normal.
   Choose **Stat**>**Basic Statistics**>**Normality Test**. Place Heights in the **V**ariable: text box. Place Leaf Spring Heights in the **T**itle: text box. Choose **OK.**

3.35 Farnum (1994, p 195) discusses a chrome plating process. Small electric currents are run through a chemical plate containing nickel resulting in a thin plating of the metal on the part. Since the bath loses nickel as the plating proceeds, the operators periodically add more nickel to the bath. The process runs 3 shifts per day. The following are the bath concentrations at the beginning of the day for 5 days. Assume that these concentrations form a random sample.

   4.8   4.5   4.4   4.2   4.4

   a. Find the sample mean, sample variance, and sample standard deviation for these data.
   b. Assume that the true mean nickel concentration of 4.5 oz/gal, which is the operating standard. Calculate the observed value for the t-statistic. Use the rough rules of thumb to interpret your result.
   c. What did you assume in order to do the above informal analysis? Can you evaluate how well these data meet these assumptions? If yes, determine how comfortable you are with them. If not, explain why.
   d. If you could remove any single observation from this data set, which one has the most influence on the sample variance? Which one has the least influence? Justify your answer statistically. (Hint: what is the definitional formula for the sample variance?)

3.36 (EXO354) Albin (1990) studied aluminum contamination in recycled PET plastic from a pilot plant operation at Rutgers University. She collected 26 samples and measured, in parts per million (ppm) the amount of aluminum contamination. The maximum acceptable level of aluminum contamination, on the average, is 220 ppm. The data follow.

291	222	125	79	145	119	244	118	182
63	30	140	101	102	87	183	60	191
119	511	120	172	70	30	90	115	

a. Find the sample mean, sample variance, and sample standard deviation for these data.

   Follow these steps to find the sample statistics.

Step 1. Enter data.

   Enter the data in column 1. Name column 1 as *PPM*.

Step 2. Obtain descriptive statistics.

   Choose **Stat**>**Basic Statistics**>**Descriptive Statistics**. Place PPM in the Variables: text box. Choose **OK**.

b. Assume that the true mean amount of aluminum contamination is 220 ppm, which is the maximum acceptable level. Calculate the observed value for the t-statistic. Use the rough rules of thumb to interpret your result.

   Follow this step to calculate the t-statistic.

   Choose **Stat**>**Basic Statistics**>**1-Sample t.** Place PPM in the Variables: text box. Darken the **T**est mean: option button. Place 220 in the **T**est mean: text box. Select the option of greater than from the Alternative drop down dialog box. Choose **OK**.

c. What did you assume in order to do the above informal analysis? Can you evaluate how well these data meet these assumptions? If yes, determine how comfortable you are with them. If not, explain why.

   Follow this step to construct a normal probability plot to examine the assumption that the assumption that the distribution of the leaf spring heights from which these observations were drawn is normal.

   Choose **Stat**>**Basic Statistics**>**Normality Test**. Place Heights in the **V**ariable: text box. Place Leaf Spring Heights in the **T**itle: text box. Choose **OK**.

d. If you could remove any single observation from this data set, which one has the most influence on the sample variance? Which one has the least influence? Justify your answer statistically. (Hint: what is the definitional formula for the sample variance?)

Chapter 4
Estimation and Testing

4.1   Overview of Estimation

Minitab may be used in estimation and testing. It is of considerable importance to be able to construct confidence intervals to estimate the parameters in a model and to address "interesting questions" about the parameters in the model. After reading this chapter you should be able to

- Construct a Confidence Interval on a Population Mean.
- Perform One Sample t-tests for a Single Mean.
- Perform Two-Sample t-tests on Independent Samples.
- Construct a Confidence Interval on the Difference Between Two Means.
- Perform a Paired t-test for Dependent Samples.

New Minitab Commands (and some Minitab commands used previously)

1. **Calc>Calculator** - Performs arithmetic using an algebraic expression. You can use arithmetic operations, comparison operations, logical operations, functions, and column operations. In this section, you will use this calculator to perform a logical operation.

2. **Stat>Tables>Tally** - Prints a one-way table of counts and percents for specified variables. Minitab displays summary information for each distinct value in the column. In this section, you will construct a one-way table of counts.

Example 4.1 - Filling Milk Cartons
     Maxcy and Lowry (1985) report on a packaging process for 8 oz. (245 gram) milk cartons. Historically, the standard deviation for these weights has been 1.65 gram. The operators monitor this process by weighing five cartons of milk per day. The weights in grams for one days weighing were:

$$263.9 \quad 266.2 \quad 266.3 \quad 266.8 \quad 265$$

Follow these steps to construct a 95% confidence interval on the population mean:
Step 1. Enter Data.
          Enter the data into column C1. Name column C1 as *Weights*.
Step 2. Construct the confidence interval.
          Choose **Stat>Basic Statistics>1-Sample-z**. Enter Weights in the **V**ariables: text box. Darken the **C**onfidence interval option button. Place 95.0 in the **C**onfidence interval **L**evel: text box. Place 1.65 in the **S**igma: text box.

Choose **OK**.

The Minitab Output

Confidence Intervals

The assumed sigma = 1.65

Variable	N	Mean	StDev	SE Mean	95.0 % CI
Weights	5	265.640	1.176	0.738	(264.194, 267.086)

Figure 4.1

The Minitab output, as shown in Figure 4.1, indicates that we may feel reasonably comfortable that the true current mean amount of milk delivered to each carton is somewhere between 264.194 and 267.086 g. This interval was actually generated by a process that actually contains the true current mean 95% of the time. We are thus 95% confident that the interval 264.194 to 267.086 g really does contain the true current mean amount, and we believe that the true mean amount is somewhere in this interval.

Computer Exercises

4.1 Use Minitab to illustrate confidence intervals when sampling from a normal distribution.

a. Use Minitab to generate 1000 random samples of <u>size 4</u> from a normal distribution with a mean of 6 and a variance of 12. Calculate the sample mean and a 95% confidence interval for each sample. Count the number of samples which fail to contain the true mean value. Comment on your results.

Follow these steps illustrate 95% confidence intervals from a simulation of the results of 1000 random samples.

Step 1. Enter random data.

Choose **Calc**>**Random Data**>**Normal**. Place 1000 in the Generate rows of data text box. Type *C1-C4* in the <u>S</u>tore in column(s): text box. Place 6 in the <u>M</u>ean: text box and $\sqrt{12} = 3.4641$ in the S<u>t</u>andard deviation: text box. Choose **OK**.

Step 2. Calculate the means(s).

Choose **Calc**>**Ro<u>w</u> Statistics**. Darken the <u>M</u>ean option button. Place C1-C4 in the Input <u>v</u>ariables: text box. Type *Mean* in the Store result in: text box. Choose **OK**.

Step 3. Calculate the confidence limit(s) using the equation $\overline{y} \pm z_{\frac{\alpha}{2}} \frac{\sigma}{\sqrt{n}}$.

Calculate the lower confidence limit(s).

Choose **Calc**>**Calculator**. Place LCL in the <u>S</u>tore result in variable: text box. Type *(Mean-1.96*3.4641/2)* in the <u>E</u>xpression: text box. Choose **OK**. (Recall the $z_{\frac{\alpha}{2}} = 1.96$ for a 95% confidence interval, $\sigma = \sqrt{12} = 3.4641$ and $\sqrt{n} = \sqrt{4} = 2$.)

Calculate the upper confidence limit(s).
Choose **Calc**>**Calculator**. Place UCL in the Store result in variable: text box. Type *(Mean+1.96*3.4641/2)* in the Expression: text box. Choose **OK**

Step 4. Determine the number of samples that contain the true mean value.

List the number of samples that contain 6.
Choose **Calc**>**Calculator**. Type *IN* in the Store result in variable: text box. Type *LCL < 6 And UCL > 6* in the Expression: text box. Choose **OK**.

Count the number of samples that contain 6.
Choose **Stat**>**Tables**>**Tally.** Place IN in the Variables: text box. Place a check in the Counts checkbox. Choose **OK**. Look at the Session window to see the count of the total number of samples that contain the true mean value (1's) and the count of the total number of samples that fail to contain the true mean value (0's).

b. Follow the previous instructions using Minitab to generate 1000 random samples of size 10 from a normal distribution with a mean of 6 and a variance of 12. Calculate the sample mean and a 95% confidence interval for each sample. Count the number of samples which fail to contain the true mean value. Comment on your results.

c. Follow the previous instructions using Minitab to generate 1000 random samples of size 30 from a normal distribution with a mean of 6 and a variance of 12. Calculate the sample mean and a 95% confidence interval for each sample. Count the number of samples which fail to contain the true mean value. Comment on your results.

4.2 Repeat the instructions for the problem 4.1, but this time sampling from a uniform distribution instead of a normal distribution.

4.3 Repeat the instructions for the problem 4.1, but this time sampling from a χ^2 distribution instead of a normal distribution.

4.2   Overview of Hypothesis Testing

New Minitab Commands

1. **Manip > Copy Columns** -Copies data from columns in the current worksheet to new columns, including all rows or a specified subset. In this section, you will use this command to select elements from one column.

2. **Calc**>**Column Statistics** - Calculates various statistics on the column you select, displaying the results and optionally storing them in a constant. In this section, you will use this command to obtain particular information about one variable in the Data window.

Under most conditions it is either impossible or unrealistic to study an entire population to obtain the value of the population parameter of interest. Statisticians have developed techniques that enable us to draw inferences about population parameters from sample statistics. Two indispensable statistical decision-making tools are:

confidence intervals to estimate the value of a parameter, and

hypothesis tests to investigate theories concerning parameters.

Minitab may be used to make inferences about the value of a population parameter.

Example 4.2 - Alpha and Chance - Filling Milk Cartons - Revisted

Consider the last example in more detail. To examine the relationship between the level of significance (α) and pure chance, let us assume that a sample of 16 milk cartons are chosen at random from a large population and their weights are measured. The population of weights is assumed to approximately normally distributed with $\mu = 245$ gram and $\sigma = 1.65$ gram. Test the hypothesis $H_0 : \mu = 245$ against the alternative $H_1 : \mu \neq 245$, using $\alpha = .05$.

Follow these steps to simulate the results of running this test 20 times.

Step 1. Enter random data.

Choose **Calc**>**Random Data**>**Normal**. Place 20 in the Generate rows of data text box. Type *C1-C16* in the Store in column(s): text box. Place 245 in the Mean: text box and 1.65 in the Standard deviation: text box. Choose **OK**.

Step 2. Calculate the row mean(s).

Choose **Calc**>**Row Statistics**. Darken the Mean option button. Place C1-C16 in the Input variables: text box. Type *Mean* in the Store result in: text box. Choose **OK**.

Step 3. Calculate the test statistic(s).

Choose **Calc**>**Calculator**. Type *ZScore* in the Store result in variable: text box. Type *((Mean - 245)/(1.65/Sqrt(16)))* in the Expression: text box. Choose **OK**.

Step 4. Determine the number of z-scores outside the interval from -1.96 to 1.96.

List the z-scores outside the interval from -1.96 to 1.96.

Choose **Manip**>**Copy Columns**. Place ZScore in the Copy from columns: text box. Type *Outside* in the To columns: text box. Select the Use Rows option button. Darken the Use rows with numeric column option button. Place ZScore in the Use rows with numeric column text box. Type -500:-1.96 1.96:500 in the Use rows with numeric column equal to: text box. Choose **OK**. Choose **OK**.

Count the number of z-scores outside the interval from -1.96 to 1.96.

Choose **Calc**>**Column Statistics**. Darken the N total option button. Place Outside in the Input variable: text box. Choose **OK**. Look at the Session window to see the count of the total number of observation in Code and at the Data window to see the actual ZScores outside the interval from -1.96 to 1.96.

The Minitab Data Window

C17	C18	C19
Mean	ZScore	Outside
245.608	1.47365	-2.11899
245.587	1.42212	
...	...	
244.126	-2.11899	
...	...	

In how many tests did you reject H_0 ? That is, how many times did you make the "incorrect decision"? In this simulation, the incorrect decision was made 1 out of 20 times or 5% of the time. In other words, a Type I error was committed 1 of the 20 times the test was conducted. Identify the zscore of each test. Are they all the same? Suppose $\alpha = .10$ was used? Does this change any of your decisions to reject or not reject the null hypothesis?

Beta and Chance

Beta is the probability of failing to reject the null hypothesis when the null hypothesis is false. If the null hypotheses is $H_0 : \mu = 245$ vs. the alternative $H_1 : \mu \neq 245$, using $\alpha = .05$, then the null hypothesis would be rejected if $|Z| > 1.96$. That would imply that the null hypothesis will be rejected for means greater than 245.8085 or less than 244.1915.

Follow these steps to simulate the results of running this test 20 times.

Step 1. Enter random data.

Choose **Calc**>**Random Data**>**Normal**. Place 20 in the Generate rows of data text box. Type *C1-C16* in the Store in column(s): text box. Place 246 (a value greater than 245.8085!!) in the Mean: text box and 1.65 in the Standard deviation: text box. Choose **OK**.

Step 2. Calculate the row mean(s).

Choose **Calc**>**Row Statistics**. Darken the Mean option button. Place C1-C16 in the Input variables: text box. Type *Mean* in the Store result in: text box. Choose **OK**.

Step 3. Calculate the test statistic(s).

Choose **Calc**>**Calculator**. Place ZScore in the Store result in variable: text box. Type *((Mean - 245)/(1.65/Sqrt(16)))* in the Expression: text box. Choose **OK**.

Step 4. Determine the number of z-scores in the interval from -1.96 to +1.96.

List the z-scores inside the interval from -1.96 to 1.96.
Choose **Manip**>**Copy Columns**. Place ZScore in the Copy from columns: text box. Type *InInterval* in the To columns: text box. Select the Use Rows option button. Darken the Use rows with numeric column option button. Place ZScore in the Use rows with numeric column text box. Place -1.96:1.96 in the Use rows with column equal to: text box. Choose **OK**. Choose **OK**.

Count the number of z-scores inside the interval from -1.96 to 1.96.

Choose $\underline{C}$alc>$\underline{C}$olumn Statistics. Darken the N $\underline{t}$otal option button. Place InInterval in the Input $\underline{v}$ariable: text box. Place Count in the Store result in: text box. Choose $\underline{OK}$.

The Minitab Data Window

C17	C18	C19
Mean	ZScore	InInterval
246.055	2.55741	0.67301
245.278	0.67301	
...	...	.
246.181	2.86199	
...	...	

In how many tests did you to fail to reject H_0 ? That is, how many times did you make the "incorrect decision"? In this simulation, the incorrect decision was made 4 out of 20 times or 20% of the time. In other words, an incorrect decision was made 4 out of 20 times and thus a Type II error was committed. How would this example change if a mean further away from 245 were chosen?

Computer Exercises

4.4 Use Minitab to illustrate Type I and Type II error rates for the hypotheses

$$H_0 \quad : \quad \mu = 6$$
$$H_0 \quad : \quad \mu > 6$$

when sampling from a normal distribution with a variance known to be 12. In this situation, the appropriate test statistic is

$$Z = \frac{\bar{y} - 6}{\sqrt{\frac{12}{n}}}.$$

Suppose we wish to use a .05 significance level for our test. We thus should reject the null hypothesis whenever $Z > 1.645$.

a. Use Minitab to generate 1000 random samples of size 4 from a normal distribution with a mean of 6 and a variance of 12, which corresponds to the null hypothesis. Calculate the sample mean and Z for each sample. Count the number of samples which yield values of $Z > 1.645$, i.e. which reject the null hypothesis. Comment on your results.

Follow these steps illustrate Type I error rates from a simulation of the results of running this test 1000 times.

Step 1. Enter random data.

Choose $\underline{C}$alc>$\underline{R}$andom Data>$\underline{N}$ormal. Place 1000 in the Generate rows of data text box. Type *C1-C4* in the $\underline{S}$tore in column(s): text box. Place 6 in the $\underline{M}$ean: text box and $\sqrt{12} = 3.4641$ in the Standard deviation: text box. Choose $\underline{OK}$.

Step 2. Calculate the row mean(s).

Choose $\underline{C}$alc>$\underline{Row}$ Statistics. Darken the $\underline{M}$ean option button. Place C1-C4 in the Input $\underline{v}$ariables: text box. Type *Mean* in the Store result in: text box. Choose $\underline{OK}$.

Step 3. Calculate the test statistic(s).

Choose **Calc**>**Calculator.** Type *ZScore* in the Store result in variable: text box. Place

((Mean - 6)/(3.4641/Sqrt(4))) in the Expression: text box. Choose **OK**.

Step 4. Count the number of z-scores greater than 1.645.

List the z-scores greater than 1.645.

Choose **Manip**>**Copy Columns.** Place ZScore in the Copy from columns: text box. Type Outside in the To columns: text box. Select the Use Rows option button. Darken the Use rows with numeric column option button. Place ZScore in the Use rows with numeric column text box. Place 1.645:500 in the Use rows with numeric column equal to: text box. Choose **OK**. Choose **OK**.

Count the number of z-scores greater than 1.645.

Choose **Calc**>**Column Statistics**. Darken the N total option button. Place Outside in the Input variable: text box. Choose **OK**. Look at the Session window to see the count of the total number of z-scores greater than 1.645.

Follow these steps illustrate Type II error rates from a simulation of the results of running this test 1000 times.

Step 1. Enter random data.

Choose **Calc**>**Random Data**>**Normal**. Place 1000 in the Generate rows of data text box. Type *C1-C4* in the Store in column(s): text box. Place 9 in the Mean: text box and $\sqrt{12} = 3.4641$ in the Standard deviation: text box. Choose **OK**.

Step 2. Calculate the row mean(s).

Choose **Calc**>**Row Statistics**. Darken the Mean option button. Place C1-C4 in the Input variables: text box. Type *Mean* in the Store result in: text box. Choose **OK**.

Step 3. Calculate the test statistic(s).

Choose **Calc**>**Calculator.** Type *ZScore* in the Store result in variable: text box. Type

((Mean - 6)/(3.4641/Sqrt(4))) in the Expression: text box. Choose **OK**.

Step 4. Count the number of z-scores less than 1.645.

List the z-scores less than 1.645.

Choose **Manip**>**Copy Columns.** Place ZScore in the Copy from columns: text box. Place Code in the To columns: text box. Select the Use Rows option button. Darken the Use rows with numeric column option button. Place ZScore in the Use rows with numeric column text box. Place -500:1.645 in the Use rows with numeric column equal to: text box. Choose **OK**. Choose **OK**.

Count the number of z-scores less than 1.645.

Choose **Calc**>**Column Statistics**. Darken the N total option button. Place Code in the Input variable: text box. Choose **OK**. Look at the Session window to see the count of the total number of observation in Code.

b. Follow the previous instructions using Minitab to generate 1000 random samples of size 4 from a normal distribution with a mean of 8 and a variance of 12. Calculate the sample mean and Z for each sample. Count the number of samples which yield values of Z > 1.645, i. e. which reject the null hypothesis. Comment on your results.

c. Follow the previous instructions using Minitab to generate 1000 random samples of size 4 from a normal distribution with a mean of 10 and a variance of 12. Calculate the sample mean and Z for each sample. Count the number of samples which yield values of Z > 1.645, i.e. which reject the null hypothesis. Comment on your results.

d. Follow the previous instructions using Minitab to generate 1000 random samples of size 16 from a normal distribution with a mean of 6 and a variance of 12, which corresponds to the null hypothesis. Calculate the sample mean and Z for each sample. Count the number of samples which yield values of Z > 1.645, i.e. which reject the null hypothesis. Comment on your results.

e. Follow the previous instructions using Minitab to generate 1000 random samples of size 16 from a normal distribution with a mean of 8 and a variance of 12. Calculate the sample mean and Z for each sample. Count the number of samples which yield values of Z > 1.645, i.e. which reject the null hypothesis. Comment of your results.

f. Follow the previous instructions using Minitab to generate 1000 random samples of size 16 from a normal distribution with a mean of 10 and a variance of 12. Calculate the sample mean and Z for each sample. Count the number of samples which yield values of Z > 1.645, i. e. which reject the null hypothesis. Comment of your results.

4.5 Repeat the instructions for the problem 4.4, but this time sampling from a uniform distribution over the interval 0 to 12, instead of a normal distribution.

4.3   Inference for a Single Mean

New Minitab Commands

1. **Stat**>**Basic Statistics**>**1-Sample t** - Performs a one sample t-test or t-confidence interval for the mean. In this section, you will use this command to perform one sample t-tests and calculate confidence intervals.

One-Sided t-Tests

In most situations the population standard deviation, σ, is unknown. Minitab can use Student's t test to make inferences about the value of the population parameter μ when σ is unknown. This procedure may be applied to samples of all sizes where the assumption is that the parent population is approximately normally distributed.

Example 4.3 - Porosities of Battery Plates

Nickel-Hydrogen (Ni-H) batteries use a nickel plate as its anode. The sintering process, whereby the plates are "fired" at high temperature, essentially controls the plate's porosity. The manufacturer has set a target porosity of 80%. Determine if the mean porosity is less than 80% at the .05 significance level based upon the following data.

$$
\begin{array}{ccccc}
79.1 & 79.5 & 79.3 & 79.3 & 78.8 \\
79.0 & 79.2 & 79.7 & 79.0 & 79.2
\end{array}
$$

Follow these steps to perform the hypothesis test.

Step 1. Enter data.

Enter the observations into column C1. Name column C1 as *Porosity*.

Step 2. Calculate the test statistic.

Choose **Stat**>**Basic Statistics**>**1-Sample t**. Place Porosity in the **V**ariables: text box. Darken the **T**est mean: option button. Place 80 in the **T**est mean: text box. Choose less than from the **A**lternative: drop down dialog box. Choose **OK**.

The Minitab Output

T-Test of the Mean

```
Test of mu = 80.0000 vs mu < 80.0000

Variable    N      Mean    StDev   SE Mean      T        P
Porosity   10   79.2100   0.2601   0.0823   -9.60   0.0000
```

Figure 4.2

The Minitab output, as shown in Figure 4.2, indicates the descriptive statistics (N, Mean, StDev, SE Mean), as well as the observed t value of -9.60, and the p-value of 0.0000 (meaning $p < 0.001$). Since p is less than α, the null hypothesis $H_0 : \mu = 80$ is rejected. We thus have evidence to suggest that the true mean porosity is less than 80%, which supports the contention that the sintering process is overfiring the plates.

The Two-Sided t-Test

Example 4.4 - Grinding of Silicon Wafers for Integrated Circuits

Roes and Does (1995) present data on the grinding of silicon wafers used in integrated circuits. The target is $244\mu m$. Determine if the mean thickness is different from $244\mu m$ at the .01 significance level based upon the following data.

$$240 \quad 243 \quad 250 \quad 253 \quad 248$$

Follow these steps to perform the hypothesis test.

Step 1. Enter data.

Enter the observations into column C1. Name column C1 as *Thickness*.

Step 2. Calculate the test statistic.

Choose **Stat**>**Basic Statistics**>**1-Sample t**. Place Thckness in the **V**ariables: text box. Darken the option button for **T**est mean:. Place 244 in the **T**est mean: text box. Choose the (default) option of not equal in the **A**lternative: drop down dialog box. Choose **OK**.

The Minitab Output

T-Test of the Mean

```
Test of mu = 244.00 vs mu not = 244.00

Variable    N     Mean   StDev  SE Mean     T     P
Thicknes    5   246.80    5.26     2.35  1.19  0.30
```

Figure 4.3

The Minitab output, as shown in Figure 4.3, indicates the descriptive statistics (N, Mean, StDev, SE Mean), as well as the observed t value of 1.19 (which is not greater than $|t_{4,.005}| > 4.604$), and the p-value of 0.30 (meaning $p > 0.01$). Since p is greater than α, the null hypothesis $H_0 : \mu = 244$ is not rejected. We thus do not have sufficient evidence to suggest that the thickness is significantly different from 244μm.

Confidence Intervals for the Mean

Follow this step to construct a 99% confidence interval on the population mean thickness of the silicon wafers:

Choose **Stat**>**Basic Statistics**>**1-Sample t**. Enter Thickness in the **V**ariables: text box. Darken the **C**onfidence interval option button. Place 99.0 in the **C**onfidence interval **L**evel: text box. Choose **OK** .

The Minitab Output

Confidence Intervals

```
Variable    N     Mean   StDev  SE Mean       99.0 % CI
Thicknes    5   246.80    5.26     2.35  ( 235.96,  257.64)
```

Figure 4.4

The Minitab output, as shown in Figure 4.4, indicates that 99% of the sample means for a sample of size five will fall in the interval from 235.96μm to 257.64μm. Thus the confidence interval is consistent with the results of the hypothesis test.

Exercises

4.6 (EXO415) Yashchin (1992) studied the thicknesses of metal wires produced in a chip-manufacturing process. Ideally, these wires should have a target thickness of 8 microns. The data, in microns, follow.

8.4	8.0	7.8	8.0	7.9	7.7	8.0	7.9	8.2	7.9
7.9	8.2	7.9	7.8	7.9	7.9	8.0	8.0	7.6	8.2
8.1	8.1	8.0	8.0	8.3	7.8	8.2	8.3	8.0	8.0
7.8	7.9	8.4	7.7	8.0	7.9	8.0	7.7	7.7	7.8
7.8	8.2	7.7	8.3	7.8	8.3	7.8	8.0	8.2	7.8

a. Conduct the most appropriate hypothesis test using a .05 significance level.

Follow these steps to perform the hypothesis test.

Step 1. Enter data.
Enter the observations into column C1. Name column C1 as *Thickness*.

Step 2. Calculate the test statistic.
Choose **Stat**>**Basic Statistics**>**1-Sample t**. Place Thickness in the Variables: text box. Darken the option button for Test mean:. Place 8 in the Test mean: text box. Choose the appropriate option in the Alternative: drop down dialog box. Choose **OK**.

b. Construct a 95% confidence interval for the true mean thickness.
Follow this step to construct a 95% confidence interval on the population mean thickness of the metal wires.
Choose **Stat**>**Basic Statistics**>**1-Sample t**. Enter Thickness in the Variables: text box. Darken the Confidence interval option button. Place 95.0 in the Confidence interval Level: text box. Choose **OK** .

c. What did you assume in order to do the above analyses? Can you evaluate how well these data meet these assumptions? If yes, determine how comfortable you are with them. If not, explain why.

4.7 (EXO416) Albin (1990) studied aluminum contamination in recycled PET plastic from a pilot plant operation at Rutgers University. She collected 26 samples and measured, in parts per million (ppm) the amount of aluminum contamination. The maximum acceptable level of aluminum contamination, on the average, is 220 ppm. The data follow.

291	222	125	79	145	119	244	118	182
63	30	140	101	102	87	183	60	191
119	511	120	172	70	30	90	115	

a. Conduct the most appropriate hypothesis test using a .01 significance level.
b. Construct a 99% confidence interval for the true mean concentration.
c. What did you assume in order to do the above analyses? Can you eval-

uate how well these data meet these assumptions? If yes, determine how comfortable you are with them. If not, explain why.

4.8 Montgomery (1991 p 266) reports results for a process which manufactures high-voltage supplies which have a nominal output voltage of 350 V. The production people are concerned that the process is beginning to produce power supplies with a true mean output voltage somewhat larger than the nominal. Following are the data for the last four power supplies tested.

351.4   351.5   351.2   351.6

    a. Conduct the most appropriate hypothesis test using a .10 significance level.

    b. Construct a 90% confidence interval for the true mean voltage.

    c. What did you assume in order to do the above analyses? Can you evaluate how well these data meet these assumptions? If yes, determine how comfortable you are with them. If not, explain why.

4.9 Farnum (1994, p 195) discusses a chrome plating process. Small electric currents are run through a chemical plate containing nickel resulting in a thin plating of the metal on the part. Since the bath loses nickel as the plating proceeds, the operators periodically add more nickel to the bath.. Operating standards call for a nickel concentration of 4.5 oz/gal. The process runs 3 shifts per day. The following are the bath concentrations at the beginning of the day for 5 days. Assume that these concentrations form a random sample.

4.8   4.5   4.4   4.2   4.4

    a. Conduct the most appropriate hypothesis test using a .05 significance level.

    b. Construct a 95% confidence interval for the true mean concentration.

    c. What did you assume in order to do the above analyses? Can you evaluate how well these data meet these assumptions? If yes, determine how comfortable you are with them. If not, explain why.

4.10 (EXO419) DeVor, Chang, and Sutherland (1992, pp 406-407) discuss the production of polyol, which is reacted with isocynate in a foam molding process. Variations in the moisture content of polyol causes problems in controlling the reaction with isocynate. Production has set a target moisture content of 2.125%. the following data represent 27 moisture analyses over a four month period.

2.29	2.22	1.94	1.90	2.15	2.02	2.15	2.09	2.18
2.00	2.06	2.02	2.15	2.17	2.17	1.90	1.72	1.75
2.12	2.06	2.00	1.98	1.98	2.02	2.14	2.10	2.05

    a. Conduct the most appropriate hypothesis test using a .01 significance level.

    Follow these steps to perform the hypothesis test.

    Step 1. Enter data.

            Enter the observations into column C1. Name column C1 as *Moisture*.

    Step 2. Calculate the test statistic.

Choose **Stat**>**Basic Statistics**>**1-Sample t**. Place Moisture in the Variables: text box. Darken the option button for Test mean:. Place 2.125 in the Test mean: text box. Choose the appropriate option in the Alternative: drop down dialog box. Choose **OK**.

b. Construct a 99% confidence interval for the true mean moisture content.
Follow this step to construct a 99% confidence interval on the population mean moisture content.
Choose **Stat**>**Basic Statistics**>**1-Sample t**. Enter Moisture in the Variables: text box. Darken the Confidence interval option button. Place 99.0 in the Confidence interval Level: text box. Choose **OK** .

c. What did you assume in order to do the above analyses? Can you evaluate how well these data meet these assumptions? If yes, determine how comfortable you are with them. If not, explain why.

4.11 (EXO420) Weaver (1990) examined a galvanized coating process for large pipes. Standards call for an average coating weight of 200 pounds per pipe. The following data are the coating weights for a random sample of 30 pipes.

216 202 208 208 212 202 193 208 206 206
206 213 204 204 204 218 204 198 207 218
204 212 212 205 203 196 216 200 215 202

a. Conduct the most appropriate hypothesis test using a .01 significance level.
b. Construct a 99% confidence interval for the true mean coating weight.
c. What did you assume in order to do the above analyses? Can you evaluate how well these data meet these assumptions? If yes, determine how comfortable you are with them. If not, explain why.

4.12 Holmes and Mergen (1992) studied a batch operation at a chemical plant where an important quality characteristic was the product viscosity with a target value of 14.90. Production personnel use a viscosity measurement for each 12 hour batch to monitor this process. The viscosities for the past 10 batches follow.

13.3 14.5 15.3 15.3 14.3
14.8 15.2 14.9 14.6 14.1

a. Conduct the most appropriate hypothesis test using a .10 significance level.
b. Construct a 90% confidence interval for the true mean visosity.
c. What did you assume in order to do the above analyses? Can you evaluate how well these data meet these assumptions? If yes, determine how comfortable you are with them. If not, explain why.

4.13 (EXO422) McNeese and Klein (1991) looked at the average particle size of a product with a specification of 70 - 130 microns and a target of 100 microns. Production personnel measure the particle size distribution using a set of screening sieves. They test one sample a day in order to monitor this process.

The average particle sizes for the past 25 days follow.

99.6	92.1	103.8	95.3	101.6
102.8	100.9	100.5	102.7	96.9
101.5	96.7	96.8	97.8	104.7
103.2	97.5	98.3	105.8	100.6
102.3	93.8	102.7	94.9	94.9

a. Conduct the most appropriate hypothesis test using a .05 significance level.

Follow these steps to perform the hypothesis test.

Step 1. Enter data.

Enter the observations into column C1. Name column C1 as *Particle*.

Step 2. Calculate the test statistic.

Choose **Stat**>**Basic Statistics**>**1-Sample t**. Place Particle in the Variables: text box. Darken the option button for Test mean:. Place 100 in the Test mean: text box. Choose the appropriate option in the Alternative: drop down dialog box. Choose **OK**.

b. Construct a 95% confidence interval for the true mean particle size.
Follow this step to construct a 99% confidence interval on the population mean particle size.
Choose **Stat**>**Basic Statistics**>**1-Sample t**. Enter Particle in the Variables: text box. Darken the Confidence interval option button. Place 95.0 in the Confidence interval Level: text box. Choose **OK** .

b. What did you assume in order to do the above analyses? Can you evaluate how well these data meet these assumptions? If yes, determine how comfortable you are with them. If not, explain why.

4.4   Inferences for Two Independent Samples

New Minitab Commands

1. **Stat**>**Basic Statistics**>**2-Sample t** - Performs an independent two-sample t-test and generates a confidence interval.

   a. **Samples in one column** - Choose if the groups are stacked in the same column, differentiated by subscript values (group codes) in a second column. In this section, you will use this command to perform a two-sample t-test where all the data is in one column, with the subscripts in a second column.

   b. **Samples in different columns** - Choose if the groups are in two separate columns. You will not use this command, in this section, to perform a two-sample t-test where the data is in two seperate columns.

2. **Graph**>**Boxplot** - Produces a boxplot (also called box-and-whisker plot). A

default boxplot consists of a box, whiskers, and outliers. Minitab draws a line across the box at the median. In this section, you will use this command to construct parallel boxplots.

Up to this point we have been interested in a single parameter of a single population. More typically, we are interested in the characteristics of two or more distinct populations. Assumptions are that the observations are approximately normally distributed, the two random samples are independent, and the variances for each sample estimate a common population variance.

Example 4.5 - Packaging of Ground Beef

Maxcy and Lowry (1985) report on a packaging process for ground beef over a series of days. The question of interest is whether the true mean amount delivered by this process changes from day to day. Ten packages, which are accurately weighed, are randomly selected over the course of the day for two sequential days. A .05 level of significance is chosen for the hypothesis test of $H_0 : \mu_1 - \mu_2 = 0$ against $H_0 : \mu_1 - \mu_2 \neq 0$. The weights are as follows.

First Day

1397.8	1394.8	1391.7	1400.0	1393.5
1391.2	1384.0	1391.0	1385.7	1385.3

Second Day

1410.0	1393.9	1405.9	1404.2	1387.3
1398.5	1399.9	1392.5	1402.5	1391.8

Follow these steps to perform the hypothesis test.

Step 1. Enter data.

Enter the observations for the the first day into column C1 and continue to enter the observations for the second day into column C1 directly underneath the observations for the first day. Name column C1 as *Weights*. Code the first day as 1 by placing ten 1's in column C2. Code the second day as 2 by placing ten 2's directly underneath the ten 1's in column C2. Name column C2 as *Days*.

Step 2. Calculate the test statistic.

Choose **Stat**>**Basic Statistics**>**2-Sample t**. Darken the option button for Samples in <u>o</u>ne column. Place Weights in the Sa<u>m</u>ples: text box. Place Days in the Su<u>b</u>scripts: text box. Choose the (default) option of not equal in the <u>A</u>lternative: drop down dialog box. Place a check in the Assume <u>e</u>qual variances check box. Choose **OK**.

The Minitab Output

Two Sample T-Test and Confidence Interval

```
Two sample T for Weights
Days          N      Mean      StDev    SE Mean
1            10    1391.50      5.33       1.7
2            10    1398.65      7.19       2.3

95% CI for mu (1) - mu (2): ( -13.1,  -1.2)
T-Test mu (1) = mu (2) (vs not =): T= -2.53  P=0.021  DF=   18
Both use Pooled StDev = 6.33
```

Figure 4.5

The Minitab output, as shown in Figure 4.5, indicates a number of descriptive statistics (N, Mean, StDev, SE Mean), as well as the observed t value of -2.53 (which is in the rejection region where $\left| t_{18, \frac{\alpha}{2}=.025} \right| > 2.101$), and the two-tailed p-value of 0.021 (meaning p<0.05). Since p is less than α, the null hypothesis $H_0 : \mu_1 - \mu_2 = 0$ is rejected. We thus do have sufficient evidence to suggest that the true mean amounts of ground beef are different for the two days. The industrial engineer assigned to this process should seek to identify the causes for the difference and eliminate them.

Since we reject the null hypothesis, we need to give a range of plauible values for the true difference. The Minitab output, as shown in Figure 4.5, contains a 95% confidence interval for the true mean difference for these two days. Thus, the range of plauible values for the true difference in mean amounts is in the interval from -13.1 to -1.2 g.

Checking Assumptions

To check the assumptions are that the observations are approximately normally distributed, the two random samples are independent, and the variances for each sample estimate a common population variance, construct a parallel boxplot.

Follow this step to construct the parallel boxplot.

Choose **Graph**>**Boxplot**. Place Weights in the Graph Variables: Y category text box. Place Days in the Graph Variables: X category text box.

Transpose the boxplot to a horizontal boxplot.

Choose Options. Place a check in the Transpose X and Y checkbox. Choose **OK**. Enter a title.

Choose Annotation>Title. Place an appropriate title on line(s) 1 (and 2). Choose **OK**.

Choose **OK**.

The Minitab Output

Parallel Boxplots Comparing the
Amounts of Ground Beef Delivered in Two Days

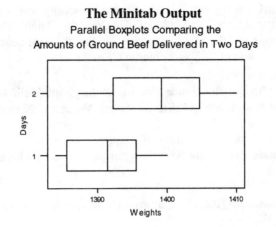

Figure 4.6

The parallel boxplots, as shown in Figure 4.6, indicate that both data sets are roughly symmetric and the tails die rapidly. The second day's amounts seme to be centered higher than the first day's. The spreads for the amounts look similar. The median value for the second day is larger than the first. The lengths of the boxes, indicating interquartile ranges, are similar indicating similar variabilities. No outliers are apparent. Since the data were taken on two different days, there is no reason todoubt the independence of the two samples. The assumptions seem to be satisfied in this case.

Exercises

4.14 (EXO429) Eibl, Kess and Pukelsheim (1992) studied the impact of viscosity upon observed coating thickness produced by a paint operation. For simplicity, they chose to study only two viscosities; "low" and "high." Up to a certain paint viscosity, higher viscosities cause thicker coatings. The engineers do not know if they have hit that limit or not. They thus wish to test whether the higher viscosity paint leads to thicker coatings. The data follow.

"Low" Viscosity

1.09	1.12	0.83	0.88	1.62	1.49	1.48	1.59
0.88	1.29	1.04	1.31	1.83	1.65	1.71	1.76

"High" Viscosity

1.46	1.51	1.59	1.40	0.74	0.98	0.79	0.83
2.05	2.17	2.36	2.12	1.51	1.46	1.42	1.40

a. **Follow these steps** to analyze these data using parallel boxplots.

Step 1. Enter data.

Enter the observations for the "Low" Viscosity into column C1 and continue to enter the observations for the "High" Viscosity into column C1 directly underneath the observations for the "Low" Viscosity. Name column C1 as *Viscosty*. Code the "Low" Viscosity observations as 1 by placing sixteen 1's in column C2. Code the "High" Viscosity observations as 2 by placing sixteen 2's directly underneath the sixteen 1's in column C2. Name column C2 as *Level*.

Step 2. Construct the boxplot.

Choose **Graph>Boxplot**. Place Viscosty in the **G**raph Variables: Y category text box. Place Level in the **G**raph Variables: X category text box.

Transpose the boxplot to a horizontal boxplot.

Choose Options. Place a check in the **T**ranspose X and Y checkbox. Choose **OK**.

Enter a title.

Choose **A**nnotation>**T**itle. Place an appropriate title on line(s) 1 (and 2). Choose **OK**.

Choose **OK**.

b. **Follow this step** to conduct the appropriate hypothesis test using a .05 significance level.

Calculate the test statistic.

Choose **Stat>Basic Statistics>2-Sample t**. Darken the option button for Samples in **o**ne column. Place Viscosty in the Sa**m**ples: text box. Place Level in the Su**b**scripts: text box. Choose the appropriate option in the **A**lternative: drop down dialog box. Place a check in the Assume **e**qual variances check box. Choose **OK**.

c. Read the Minitab output from the previous step to identify the 95% confidence interval for the true difference in the mean coating thicknesses.

d. What did you assume in order to do the above analyses? Can you evaluate how well these data meet these assumptions? If yes, determine how comfortable you are with them. If not, explain why.

e. Discuss the conclusions of the hypothesis test and confidence interval relative to the boxplot.

4.15 (EXO430) A major manufacturer of aircraft (see Montgomery 1991, pp 242-244) closely monitors the viscosity of an aircraft primer paint. The viscosities for two different time periods appear below.

Time Period 1

33.8	33.1	34.0	33.8	33.5
34.0	33.7	33.3	33.5	33.2
33.6	33.0	33.5	33.1	33.8

Time Period 2

33.5	33.3	33.4	33.3	34.7
34.8	34.6	35.0	34.8	34.5
34.7	34.3	34.6	34.5	35.0

a. Analyze these data using parallel boxplots.

b. Conduct the appropriate hypothesis test using a .05 significance level.

c. Construct a 95% confidence interval for the true difference in the mean viscosities.

d. What did you assume in order to do the above analyses? Can you evaluate how well these data meet these assumptions? If yes, determine how comfortable you are with them. If not, explain why.

e. Discuss the conclusions of the hypothesis test and confidence interval relative to the boxplot.

4.16 Galinsky et al. (1993) studied the impact of sensory modalities (either aural or visual) upon people's ability to monitor a specific display for critical events to which the observer must respond. Such tasks are critical components of such jobs as air traffic control, industrial quality control, robotic manufacturing operations, and nuclear power plant monitoring. One aspect of their study focused upon the difference in response between aural and visual stimuli. In particular, they monitored the motor activity of the subject's dominant wrist as a measure of "restlessness" or "fidgeting." The greater the activity, the more restless the subject. Galinsky and her colleagues recorded the number of wrist movements over ten minute periods of time. The data follow.

Auditory

418	236	281	416	578
329	197	397	677	698

Time Period 2

386	517	617	870	892
416	574	782	838	885

a. Analyze these data using parallel boxplots.

b. Conduct the appropriate hypothesis test using a .01 significance level.

c. Construct a 99% confidence interval for the true difference in the mean number of wrist movements.

d. What did you assume in order to do the above analyses? Can you evaluate how well these data meet these assumptions? If yes, determine how comfortable you are with them. If not, explain why.

e. Discuss the conclusions of the hypothesis test and confidence interval relative to the boxplot.

4.17 (EXO432) An independent consumers group tested tires from two major brands of radial tires to determine whether there were any differences in the expected tread life. The data, in thousands of miles, follow.

Brand 1

50	54	52	47	61
56	51	51	48	56
53	43	58	52	48

Brand 2

57	61	47	52	53
57	56	53	67	57
62	56	56	62	57

a. **Follow these steps** to analyze these data using parallel boxplots.

Step 1. Enter data.

Enter the observations for "Brand 1" into column C1 and continue to enter the observations for "Brand 2" into column C1 directly underneath the observations for "Brand 1". Name column C1 as *Tredlife*. Code the "Brand 1" observations as 1 by placing fifteen 1's in column C2. Code the "Brand 2" observations as 2 by placing fifteen 2's directly underneath the fifteen 1's in column C2. Name column C2 as *Brands*.

Step 2. Construct the boxplot.

Choose **Graph>Boxplot**. Place Tredlife in the Graph Variables: Y category text box. Place Brands in the Graph Variables: X category text box.

Transpose the boxplot to a horizontal boxplot.

Choose Options. Place a check in the Transpose X and Y checkbox. Choose **OK**.

Enter a title.

Choose Annotation>Title. Place an appropriate title on line(s) 1 (and 2). Choose **OK**. Choose **OK**.

b. **Follow this step** to conduct the appropriate hypothesis test using a .05 significance level.

Choose **Stat>Basic Statistics>2-Sample t**. Darken the option button for Samples in one column. Place Tredlife in the Samples: text box. Place Brands in the Subscripts: text box. Choose the appropriate option in the Alternative: drop down dialog box. Place a check in the Assume equal variances check box. Choose **OK**.

c. Read the Minitab output from the previous step to identify the 95% confidence interval for the true difference in the mean coating thicknesses.

d. What did you assume in order to do the above analyses? Can you evaluate how well these data meet these assumptions? If yes, determine how comfortable you are with them. If not, explain why.

e. Dicuss the conclusions of the hypothesis test and confidence interval relative to the boxplot

4.18 Nelson (1989) compared two brands of ultrasonic humidifiers with respect to the rate at which they output moisture. The following data are the m cimum outputs in fluid ounces per hour as measured in a chamber controlled at a temperature of 70^0F and a relative humidity of 30%.

Brand 1

14.0   14.3   12.2   15.1

Brand 2

12.1   13.6   11.9   11.2

a. Analyze these data using parallel boxplots.

b. Conduct the appropriate hypothesis test using a .10 significance level.

c. Construct a 90% confidence interval for the true difference in the mean viscosities.

d. What did you assume in order to do the above analyses? Can you evaluate how well these data meet these assumptions? If yes, determine how comfortable you are with them. If not, explain why.

e. Discuss the conclusions of the hypothesis test and confidence interval relative to the boxplot.

4.19 (EXO434) The following data are the yields for the last eight hours of prodcution from two ethanol-water distallation columns.

Column 1

70   74   73   72   72   73   72   73

Column 2

71   74   72   71   72   70   72   72

a. Analyze these data using parallel boxplots.

b. Conduct the appropriate hypothesis test using a .10 significance level.

c. Construct a 90% confidence interval for the true difference in the mean yields.

d. What did you assume in order to do the above analyses? Can you evaluate how well these data meet these assumptions? If yes, determine how comfortable you are with them. If not, explain why.

e. Discuss the conclusions of the hypothesis test and confidence interval relative to the boxplot.

4.5   The Paired t-Test

New Minitab Commands

1. **Calc>Calculator** - Peforms arithmetic using an algebraic expression. You can use arithmetic operations, comparison operations, logical operations, functions, and column operations. In this section, you will use this command to calculate the differences between two observations taken from two dependent

samples.

2. **Stat>Basic Statistics>1-Sample t** - Performs a one sample t-test or t-confidence interval for the mean. In this section, you will use this command to perform paired difference t-tests.

Until now, all of the two group comparisons have assumed that the two samples are independent. If two observations are taken from the sample sampling unit, the samples are said to be dependent samples.

Example 4.6 - Testing Octane Blends

Snee (1981) determined the octane ratings of 32 gasoline blends by two standard methods: motor (method 1) and research (method 2). The question of interest is whether one method tends to produce higher results than the other. The sample consists of 32 gasoline blends covering a wide range of target octane ratings. Each blend is divided into two samples so that each blend can be tested by both methods. This situtation is an example of a paired study since the data are "paired" by the particular gasoline blend and thus the samples are said to be dependent samples. A .01 level of significance will be used to test the null hypothesis $H_0 : \delta = 0$ against the alternative hypothesis $H_1 : \delta \neq 0$. The ratings are as follows.

Method

1	105.0	81.4	91.4	84.0	88.1	91.4	98.0	90.2
2	106.6	83.3	99.4	94.7	99.7	94.1	101.9	98.6

1	94.7	105.5	86.5	83.1	86.2	87.7	84.7	83.8
2	103.1	106.2	92.3	89.2	93.6	97.4	88.8	85.9

1	86.8	90.2	92.4	85.9	84.8	89.3	91.7	87.7
2	96.5	99.5	99.8	97.0	95.3	100.2	96.3	93.9

1	91.3	90.7	93.7	90.0	85.0	87.9	85.2	87.4
2	97.4	98.4	101.3	99.1	92.8	95.7	93.5	97.5

Follow these steps to perform the hypothesis test.

Step 1. Enter data.

Enter the observations for Method 1 in column C1. Name column C1 as *Method1*. Enter the observations for Method 2 in column C2. Name column C2 as *Method2*.

Step 2. Calculate the paired differences.

Choose **Calc>Calculator**. Type *Diff* in the Store result in variable: text box. Click in the Expression: text box. Use the mouse to highlight Method1 and double click (or Select). Type -. Use the mouse to highlight Method2 and double click (or Select). Choose **OK**.

Step 3. Calculate the test statistic.

Choose **Stat>Basic Statistics>1-Sample t**. Place Diff in the Variables:

text box. Darken the option button for Test mean:. Place 0 in the Test mean: text box. Choose the (default) option of not equal in the Alternative: drop down dialog box. Choose **OK**.

The Minitab Output

T-Test of the Mean

```
Test of mu = 0.000 vs mu not = 0.000

Variable    N    Mean   StDev   SE Mean        T        P
Diff       32  -7.103   3.040    0.537   -13.22   0.0000
```

Figure 4.7

The Minitab output, as shown in Figure 4.7, indicates a number of descriptive statistics (N, Mean, StDev, SE Mean), as well as the observed t value of -13.22 (which is in the rejection region where $\left|t_{31,\frac{\alpha}{2}=.025}\right| > 2.750$), and the p-value of 0.0000 (meaning p<0.01). Since p is less than α, the null hypothesis $H_0 : \delta = 0$ is rejected. We thus do have sufficient evidence to suggest that the Method 2 yields higher octane ratings than Method 1.

Since we reject the null hypothesis we need to give a range for the plauible values of the true mean differences.

Follow this step to construct a 99% confidence interval on the population mean difference in ratings.
Choose **Stat**>**Basic Statistics**>**1-Sample t**. Enter Diff in the Variables: text box. Darken the Confidence interval option button. Place 99.0 in the Confidence interval Level: text box. Choose **OK** .

The Minitab Output

Confidence Intervals

```
Variable    N   Mean   StDev   SE Mean        99.0 % CI
Diff       32 -7.103   3.040    0.537   ( -8.578,  -5.628)
```

Figure 4.8

The Minitab output, as shown in Figure 4.8, indicates that the plausible values for this difference range from -8.578 to -5.628, which again indicates that method 2 yield higher outane ratings than method 1. In particular, method 2 seems to yield ratings somewhere between 5.6 and 8.6 points higher. Thus the confidence interval is consistent with the results of the hypothesis test.

Exercises

4.20 (EXO435) Grubbs (1983) described data on the running time of 20 fuses. Two operators, acting independently, measured the time for each fuse. The data follow.

Oper.

1	4.85 4.93 4.75 4.77 4.67 4.87 4.67 4.94 4.85 4.75
2	5.09 5.04 4.95 5.02 4.90 5.05 4.90 5.15 5.08 4.98

1	4.83 4.92 4.74 4.99 4.88 4.95 4.95 4.93 4.92 4.89
2	5.04 5.12 4.95 5.23 5.07 5.23 5.16 5.11 5.11 5.08

a. Conduct the most appropriate hypothesis test using a .05 significance level.

b. Construct a 95% confidence interval for the true mean difference in time.

c. What did you assume in order to do the above analyses? Can you evaluate how well these data meet these assumptions? If yes, determine how comfortable you are with them. If not, explain why.

4.21 (EXO436) Nickel-Hydrogen batteries use nickel plates as the anode. After the plates are sintered or fired in a high temperature furnace, they are grouped into lots of 40 plates each, and then placed into an "electrode deposition" (ED) bath where they are placed under an electrical load. This bath controls the electrical properties of the cell. An important characteristic of the nickel plates batteries is "stress growth." As the battery cell undergoes its charge-discharge cycle, the plates actually begin to expand due to the stress. One of the engineers believes that the more porous the plate, the greater the stress growth. He wants to conduct a test to confirm this belief. The engineer knows that the specific conditions of the ED bath have a major impact on stress growth. Since no two ED bath runs are identical, the engineer expects a lot of variability in stress growth purely from the ED baths. To minimize the impact of the ED baths, he has set up each ED lot so that 20 plates have "low" porosity plates and 20 plates have "high" porosity plates. After the ED run, the engineer randomly selects five low porosity plates to make a test battery cell and five high porosity plates to form a second test cell. The following real data are the average per cent increases in the plates' thickness after 200 charge-discharge cycles. The following data summarize the results from 16 different ED lots.

Porosity

low	1.43 3.56 2.03 0.92 3.21 3.08 3.69 2.81
high	3.00 9.41 3.81 1.81 4.42 2.19 1.02 2.81

low	2.34 2.39 0.95 2.01 1.98 1.59 1.04 1.66
high	4.47 4.18 3.46 2.67 1.23 1.95 0.51 0.08

a. **Follow these steps** to conduct the most appropriate hypothesis test using a .05 significance level.

Step 1. Enter data.

Enter the observations for low porosity 1 in column C1. Name column C1 as *Low*. Enter the observations for high porostiy in column C2.

Name column C2 as *High*.

Step 2. Calculate the paired differences.

Choose <u>C</u>alc><u>Ca</u>lculator. Type *Diff* in the <u>S</u>tore result in variable: text box. Click in the "Expression:" text box. Use the mouse to highlight Low and double click (or Select). Type -. Use the mouse to highlight High and double click (or Select). Choose **OK**.

Step 3. Calculate the test statistic.

Choose <u>Stat</u>><u>B</u>asic **Statistics**><u>1</u>-**Sample t**. Place Diff in the <u>V</u>ariables: text box. Darken the option button for <u>T</u>est mean:. Place 0 in the <u>T</u>est mean: text box. Choose the (default) option of not equal in the <u>A</u>lternative: drop down dialog box. Choose **OK**.

b. **Follow this step** to construct a 95% confidence interval for the true mean difference in stress growth.

Choose <u>Stat</u>><u>B</u>asic **Statistics**><u>1</u>-**Sample t**. Enter Diff in the <u>V</u>ariables: text box. Darken the <u>C</u>onfidence interval option button. Place 95.0 in the <u>C</u>onfidence interval <u>L</u>evel: text box. Choose **OK** .

c. What did you assume in order to do the above analyses? Can you evaluate how well these data meet these assumptions? If yes, determine how comfortable you are with them. If not, explain why.

4.22 (EXO437) Measuring the actual dimensions of a manufactured part is a classical problem facing many different disciplines, especially mechani- cal and industrial engineers. A mechanical engineer must grapple with the thickness of nickel plates for a Nickel-Hydrogen battery. By the way the plate is made, he can consistently identify specific locations on each plate. Thus, location A on the first plate measured is the same as location A on the second plate. He believes that one specific location, A, of the plate is consistently thicker than another specific location, B. The actual thicknesses, in mm, of ten plates follow.

Location

A   31.10 31.10 30.90 30.80 32.20
B   29.75 29.75 30.15 30.80 30.20

A   30.40 29.65 29.85 29.85 30.65
B   29.75 29.75 30.15 30.80 30.20

a. Conduct the most appropriate hypothesis test using a .05 significance level.

b. Construct a 95% confidence interval for the true mean difference in thickness.

c. What did you assume in order to do the above analyses? Can you evaluate how well these data meet these assumptions? If yes, determine how comfortable you are with them. If not, explain why.

4.23 (EXO438) Eck Industries, Inc. (see Example 3.1) manufactures cast aluminum cylinder heads which are used for liquid-cooled aircraft engines. The wall thicknesses are critical, particularly in high-altitude applications. The company seeks to compare two methods for measuring these thicknesses: ultrasound (U), which is non-destructive; and sectioning(S) the heads, which obviously is destructive. Sectioning is more accurate, but ultrasound allows the company to test a part which it can ship. The company's would like to see if ultrasound gives, on the average, a different measurement. Comparisons for 18 heads follow.

Method

U	.223 .193 .218 .201 .231 .204 .228 .223 .215
S	.224 .207 .216 .204 .230 .203 .222 .225 .224

U	.223 .237 .226 .214 .213 .233 .224 .217 .210
S	.223 .226 .232 .217 .217 .237 .224 .219 .192

a. Conduct the most appropriate hypothesis test using a .05 significance level.

b. Construct a 95% confidence interval for the true mean difference in thickness.

c. What did you assume in order to do the above analyses? Can you evaluate how well these data meet these assumptions? If yes, determine how comfortable you are with them. If not, explain why.

4.24 Van Nuland (1992) daily compares two temperature instruments: one coupled to a process computer, and the other used for visual control. Ideally, these two instruments should agree. The following data are the two temperatures for a five-day period.

Day	1	2	3	4	5
Temp1	84.6	84.5	84.4	84.6	84.3
Temp2	85.2	85.1	84.9	85.3	85.0

a. **Follow these steps** to conduct the most appropriate hypothesis test using a .05 significance level.

Step 1. Enter data.

Enter the observations for Temp1 in column C1. Name column C1 as *Temp1*. Enter the observations for Temp2 in column C2. Name column C2 as *Temp2*.

Step 2. Calculate the paired differences.

Choose **Calc**>**Calculator**. Type *Diff* in the $\underline{S}$tore result in variable: text box. Click in the "Expression:" text box. Use the mouse to highlight Temp1 and double click (or Select). Type -. Use the mouse to highlight Temp2 and double click (or Select). Choose **OK**.

Step 3. Calculate the test statistic.

Choose **Stat**>**Basic Statistics**>**1-Sample t**. Place Diff in the $\underline{V}$ariables: text box. Darken the option button for $\underline{T}$est mean:. Place 0 in the $\underline{T}$est mean: text box. Choose the (default) option of not equal in the $\underline{A}$lternative: drop down dialog box. Choose **OK**.

b. **Follow this step** to construct a 95% confidence interval for the true mean difference in temperatures.

Choose **Stat**>**Basic Statistics**>**1-Sample t**. Enter Diff in the Variables: text box. Darken the Confidence interval option button. Place 95.0 in the Confidence interval Level: text box. Choose **OK** .

4.25 (EXO440) A Maintenance manager for an injection molding facility must test a new repair method which should increase the expected time between repairs. She used the new method on ten different machines. For each machine, she recorded the last time between failures prior to using the new method, which she called "Current," and the first time between failures after using the new method, which she called "New." The times, in hours, follow.

Machine	1	2	3	4	5	6	7	8	9	10
Current	155	222	346	287	115	389	183	451	140	252
New	211	345	419	274	244	420	319	505	396	222

a. Conduct the most appropriate hypothesis test using a .05 significance level.

b. Construct a 95% confidence interval for the true mean difference in temperatures.

c. What did you assume in order to do the above analyses? Can you evaluate how well these data meet these assumptions? If yes, determine how comfortable you are with them. If not, explain why.

Chapter 5
Control Charts

5.1   Overview

Processes Change over Time

Manufacturing industries use process-control ideas to improve quality and improve efficiency. Minitab can help monitor an engineering process in order to assure that important characteristics of interest remain as close to specified target values as possible by constructing control charts. After reading this chapter, you should be able to create a

- $\overline{X} - chart$ - a control chart of subgroup means.
- $\overline{R} - chart$ - a control chart of subgroup ranges.
- $\overline{X} - R\ chart$ - a control chart combining the $\overline{X} - chart$ and the $\overline{R} - chart$.
- $s - chart$ - a control chart for subgroup standard deviations.
- $x - chart$ - a control chart for individual observations.
- $np - chart$ - a control chart for the number of defectives..
- $p - chart$ - a control chart for the proportion of defectives.
- $c - chart$ - a control chart for the number of defects.

New Minitab Commands (and some old Minitab commands)

1. **Calc**>**Random Data**>**Normal** - Generates random data from a normal distribution. In this section, you will enter the population mean and the population standard deviation of a normal distribution into appropriate text boxes to simulate and experiment.

2. **Stat**>**Control Charts**>**Xbar** - Draws a control chart of subgroup means. In this section, you will use this command to create a $\overline{X} - chart$.

The Basic Idea of a Control Chart

Example 5.1 - Packaging Milk Cartons

Maxcy and Lowry (1985) report on a packaging process for 8 oz. (245 gram) milk cartons. Department of Agriculture regulations require this process to maintain a mean amount delivered of 260 g in order to ensure virtually no underfills. Historically, the standard deviation for these weights has been 1.65 gram. The operators monitor this process by weighing five cartons of milk per day. The engineer assigned to this process must devise a monitoring strategy which detects when the process begins to move away from this target mean amount. The process is considered to be in control if

$$\mu_0 - z_{\frac{\alpha}{2}} * \frac{\sigma}{\sqrt{n}} \leq \overline{y} \leq \mu_0 + z_{\frac{\alpha}{2}} * \frac{\sigma}{\sqrt{n}} \cdots$$

$\mu_0 + z_{\frac{\alpha}{2}} * \frac{\sigma}{\sqrt{n}}$ the upper control limit (UCL) and

$\mu_0 - z_{\frac{\alpha}{2}} * \frac{\sigma}{\sqrt{n}}$ the lower control limit (LCL). Typically $z_{\frac{\alpha}{2}} = 3$ is chosen. The upper and lower control limits define the expected amount of variability in the sample means if the process is in control. The graphical method using this approach is referred to as a $\overline{x}$ - chart. In this example, the target value is 260 g, the population standard deviation is 1.65 g, and a random sample of 5 cartons is taken each shift. The control limits are

$$UCL = \mu_0 + z_{\frac{\alpha}{2}} * \frac{\sigma}{\sqrt{n}} = 260 + 3 * \frac{1.65}{\sqrt{5}} = 262.2 \text{ and}$$

$$LCL = \mu_0 - z_{\frac{\alpha}{2}} * \frac{\sigma}{\sqrt{n}} = 260 - 3 * \frac{1.65}{\sqrt{5}} = 257.79.$$

The control chart, as shown in Figure 5.1, summarizes the 20 shifts studied by Maxcy and Lowry.

The Minitab Output

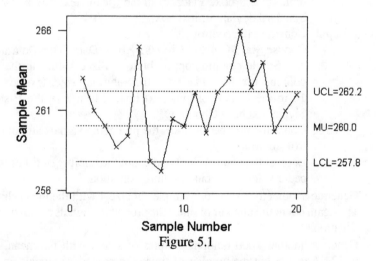

Figure 5.1

Follow these steps to simulate the experiment conducted by Maxcy and Lowry.

Step 1. Enter random data.

Choose **Calc**>**Random Data**>**Normal**. Place 100 in the Generate rows of data text box. Type *Weights* in the Store in column(s): text box. Place 260 in the **Mean**: text box and 1.65 in the Standard deviation: text box. Choose **OK**.

Step 2. Construct the control chart.

Choose **Stat**>**Control Charts**>**Xbar**. Darken the Data are arranged as a Single column: option button. Place Weights in the Single column: text box. Place 5 in the Subgroup size: text box. Place 260 in the Historical mu: text box. Place 1.65 in the Historical: Sigma text box. Choose **OK**.

Step 3. Count the number of times the procedure signals an out of control situation. Look at the control chart and count the number of times the procedure signals an out of control situation.

Computer Exercises

5.1 Consider a process which is well modeled by a normal distribution with variance of 4. Suppose the target value for the mean is 10 and that production monitors this process with an $\overline{X}$ chart using samples of size 4. The resulting upper control limit is 13, and the lower control limit is 7.

    a. Generate 1000 random samples of size 4 assuming that the process is in control. Count the number of times the procedure signals an out of control situation.

    Follow these steps to perform this simulation.

      Step 1. Enter random data.

        Choose **Calc**>**Random Data**>**Normal**. Place 4000 in the Generate rows of data text box. Type *Values* in the Store in column(s): text box. Place 10 in the Mean: text box and 2 in the Standard deviation: text box. Choose **OK**.

      Step 2. Construct the control chart.

        Choose **Stat**>**Control Charts**>**Xbar**. Darken the Data are arranged as a Single column: option button. Place Values in the Single column: text box. Place 4 in the Subgroup size: text box. Place 10 in the Historical mu: text box. Place 2 in the Historical: Sigma text box. Choose **OK**.

      Step 3. Count the number of times the procedure signals an out of control situation.

        Look at the control chart and count the number of times the procedure signals an out of control situation.

    b. Generate another 1000 random samples of size 4 with the mean shifted to 11. Again count the number of times the procedure signals an out of control situation.

    c. Generate another 1000 random samples of size 4 with the mean shifted to 12. Again count the number of times the procedure signals an out of control situation.

    d. Generate another 1000 random samples of size 4 with the mean shifted to 13. Again count the number of times the procedure signals an out of control situation.

    e. Comment on your results.

Exercises

5.1 Yashchin (1995) discusses a process for the chemical etching of silicon wafers used in integrated circuits. This process etches the layer of silicon dioxide until the layer of metal beneath is reached. This company monitors the thickness of the silicon oxide thickness since thicker layers require longer etching times. The layer has a target thickness of 1 micron and a historic standard deviation of 0.06 microns. The company uses samples of four wafers. The sample mean thicknesses for 40 samples follow. The data are in consecutive order, reading across the rows. The first observation is 1.006, the second is 1.037, etc.

> 1.006 1.037 0.944 0.957 1.012 1.035 0.917 1.067
> 1.121 0.935 0.911 1.030 1.018 0.941 1.192 1.142
> 1.138 1.188 1.080 1.228 1.153 1.141 1.179 1.190
> 1.184 0.880 0.951 0.875 0.870 0.811 0.871 0.890
> 0.866 0.794 0.868 0.854 0.905 0.885 0.885 0.977

Calculate the appropriate control limits and plot the control chart. Comment on your results.

Follow these steps to plot the control chart.

Step 1. Enter data.

Enter the data for all 40 observations in column 1. Enter 1.006 followed by 1.037, etc. Name the column *Thickness*.

Step 2. Construct the control chart.

Choose **Stat**>**Control Charts**>**Xbar**. Darken the Data are arranged as a Single column: option button. Place Thickness in the Single column: text box. Place 4 in the Subgroup size: text box. Place 1 in the Historical mu: text box. Place 0.06 in the Historical: Sigma text box. Choose **OK**.

5.2 Kane (1986) discusses the concentricity of an engine oil seal groove. Concentricity measures the cross-sectional coaxial relationship of two cylindrical features. In this case, he studied the concentricity of an oil seal groove and a base cylinder in the interior of the groove. He measures the concentricity as a positive deviation using a dial indicator gauge. Historically, this process has produced an average concentricity of 5.6 with a standard deviation of 0.7. To monitor this process, he periodically takes a random sample of three measurements. The sample mean concentricities for 20 samples follow. The data are in consecutive order, reading across the rows. The first observation is 5.48, the second is 5.83, etc.

> 5.48 5.83 6.69 6.04 5.64 5.23 4.89 5.72 5.39 6.19
> 5.78 5.76 5.82 5.63 5.73 5.05 4.62 5.74 6.60 6.81

Calculate the appropriate control limits and plot the control chart. Comment on your results.

Follow these steps to plot the control chart.

Step 1. Enter data.

Enter the data for all 20 means in column 1. Enter 5.48 followed by 5.83, etc. Name the column *Concent*.

Step 2. Construct the control chart.

Choose **Stat>Control Charts>Xbar**. Darken the Data are arranged as a Single column: option button. Place Concent in the Single column: text box. Place 1 in the Subgroup size: text box. Place 5.6 in the Historical mu: text box. Place 0.7 in the Historical: Sigma text box. Choose **OK**.

5.3 Runger and Pignatiello (1991) consider a plastic injection molding process for a part with a critical width dimension with a historic mean of 100 and historic standard deviation of 8. Periodically, clogs form in one of the feeder lines causing the mean width to change. As a result, the operator periodically takes random samples of size four. The sample mean widths for 30 samples follow. The data are in consecutive order, reading across the rows. The first observation is 93.77, the second is 105.09, etc.

93.77	105.09	106.18	103.21	97.66
103.55	90.57	105.08	95.57	102.25
100.98	101.17	103.95	100.41	101.21
100.18	105.74	90.16	103.63	102.04
105.53	112.05	112.33	119.15	109.74
106.41	112.75	106.95	105.91	115.40

Calculate the appropriate control limits and plot the control chart. Comment on your results.

5.4 Porosity is one important quality characteristic of pencil lead, which is a ceramic material "fired" at high temperatures. The porosity measures the ultimate firing state of the material. In addition, the resulting pore structure permits the lead to absorb wax in the next production step. The wax smooths the writing characteristics of the pencil. A particular pencil lead grade has a target porosity of 12.5. Historically, the porosities for this grade have had a standard deviation of 0.8. Production monitors the porosity of this grade by taking a random sample of size four from each lot of pencil lead. The sample mean porosities for 20 such lots follow. The data are in consecutive order, reading across the rows. The first observation is 12.88, the second is 12.68, etc.

12.88 12.68 12.95 11.55 13.88 13.03 13.25 12.60 13.18 12.05
12.53 12.40 12.60 12.48 12.45 12.33 12.78 12.30 11.85 11.50

Calculate the appropriate control limits and plot the control chart. Comment on your results.

5.2   Variables Control Charts

New Minitab Commands

1. **Stat**>**Control Charts**>**R** - Draws a control chart of subgroup ranges. In this section, you will use this command top construct a control chart of ranges ($R - chart$).

2. **Stat**>**Control Charts**>>**Xbar-R** - Draws a control chart for subgroup means (upper half of the screen) and a control chart for subgroup ranges (lower half of the screen) so you can examine both process level and process variation at the same time. In this section, you will construct a $\overline{X} - chart$ based on the sample range.

The Need to Monitor Both the Mean and the Variability

We monitor processes to maintain an acceptable level of quality for products. The quality of the product is typically defined through specification limits.

Example 5.2 - Grinding of Silicon Wafers

Roes and Does (1995) present data on the grinding of silicon wafers used in integrated circuits. Philips Semiconductors grinds wafers in batches of 31 and has a target of 244 μm. To monitor this process, samples of five wafers are selected from each batch. The R-Chart or Range Chart is traditionally used to monitor the process variance. The following data represent the thicknesses of 30 consecutive batches analyzed by Roes and Does.

Batch	y_{i1} y_{i2} y_{i3} y_{i4} y_{i5}	Batch	y_{i1} y_{i2} y_{i3} y_{i4} y_{i5}
1	240 243 250 253 248	16	237 239 242 247 245
2	238 242 245 251 247	17	242 244 246 251 248
3	239 242 246 250 248	18	243 245 247 252 249
4	235 237 246 249 246	19	243 245 248 251 250
5	240 241 246 247 249	20	244 246 246 250 246
6	240 243 244 248 245	21	241 239 244 250 246
7	240 243 244 249 246	22	242 245 248 251 249
8	245 250 250 247 248	23	242 245 248 243 246
9	238 240 245 248 246	24	241 244 245 249 247
10	240 242 246 249 248	25	236 239 241 246 242
11	240 243 246 250 248	26	243 246 247 252 247
12	241 245 243 247 245	27	241 243 245 248 246
13	247 245 255 250 249	28	239 240 242 243 244
14	237 239 243 247 246	29	239 240 250 252 250
15	242 244 245 248 245	30	241 243 249 255 253

Follow these steps to construct an R-chart.

Step 1. Enter data.

Enter the data for each batch in column 1. Enter the first five observations for batch 1 followed by the five observations for batch 2, etc. Name the column *Thickness*.

Step 2. Construct the control chart.

Choose <u>Stat</u>><u>Control Charts</u>><u>R</u>. Darken the Data are arranged as a Single <u>c</u>olumn: option button. Place Thickness in the Single <u>c</u>olumn: text box. Place 5 in the S<u>u</u>bgroup si<u>z</u>e: text box. Place $\frac{\overline{R}}{d_n^*} = \frac{9.25}{2.326} = 3.9768$ in the Historical: sigma text box. Choose <u>OK</u>.

The Minitab Output

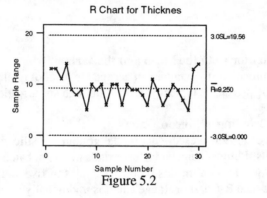

Figure 5.2

The Minitab output, as shown in Figure 5.2, contains the R-chart for all 30 samples and indicates a process which appears to be in control.

The $\overline{X}$–chart Based on the Range

Since the R-Chart indicated that the process variance is stable, a $\overline{X}$-chart can be constructed to monitor the between sample variability.

Follow these steps to construct an $\overline{X}$-chart based on the range.

Step 1. Enter data.

Enter the data for each batch in column 1. Enter the first five observations for batch 1 followed by the five observations for batch 2, etc. Name the column *Thickness*.

Step 2. Construct the $\overline{X}$-chart.

Choose <u>Stat</u>><u>Control Charts</u>><u>Xbar-R</u>. Darken the Data are arranged as a Single <u>c</u>olumn: option button. Place Thickness in the Single <u>c</u>olumn: text box. Place 5 in the S<u>u</u>bgroup si<u>z</u>e: text box. Place 245.18 in the Historical <u>m</u>ean: text box. Place $\frac{\overline{R}}{d_n^*} = \frac{9.25}{2.326} =$

3.9768   in the Historical: sigma text box. Choose **OK**.

The Minitab Output

Xbar/R Chart for Thicknes

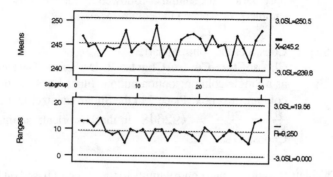

Figure 5.3

The Minitab output, as shown in Figure 5.3, contains the $\overline{X} - chart$ based on the sample range for the silicon wafer thickness data. Again, we see an in control process.

Exercises

5.5 (EXO5O7) Padgett and Spurrier (1990) analyze the breaking strengths of carbon fibers used in fibrous composite materials. These fibers measure 50 mm in length and 7-8 microns in diameter. Periodically, the manufacturer selects random samples of five fibers and tests their breaking stresses. Specifications require that 99% of the fibers must have a breaking stress of at least 1.2 GPa (giga-Pascals). The breaking stresses in GPa from 20 such samples follow.

Sample	y_{i1} y_{i2} y_{i3} y_{i4} y_{i5}	Sample	y_{i1} y_{i2} y_{i3} y_{i4} y_{i5}
1	3.7 2.7 2.7 2.5 3.6	11	1.4 3.7 3.0 1.4 1.0
2	3.1 3.3 2.9 1.5 3.1	12	2.8 4.9 3.7 1.8 1.6
3	4.4 2.4 3.2 3.2 1.7	13	3.2 1.6 0.8 5.6 1.7
4	3.3 3.1 1.8 3.2 4.9	14	1.6 2.0 1.2 1.1 1.7
5	3.8 2.4 3.0 3.0 3.4	15	2.2 1.2 5.1 2.5 1.2
6	3.0 2.5 2.7 2.9 3.2	16	3.5 2.2 1.7 1.3 4.4
7	3.4 2.8 4.2 3.3 2.6	17	1.8 0.4 3.7 2.5 0.9
8	3.3 3.3 2.9 2.6 3.6	18	1.6 2.8 4.7 2.0 1.8
9	3.2 2.4 2.6 2.6 2.4	19	1.6 1.1 2.0 1.6 2.1
10	2.8 2.8 2.2 2.8 1.9	20	1.9 2.9 2.8 2.1 3.7

a. Calculate the appropriate control limits for an R-chart and plot the data. Comment on your results.

Follow these steps to construct an R-chart.

Step 1. Enter data.

Enter the data for each sample in column 1. Enter the first five observations for sample 1 followed by the five observations for sample 2, etc. Name the column *Stresses*.

Step 2. Construct the control chart.

Choose **Stat**>**Control Charts**>**R**. Darken the Data are arranged as a Single column: option button. Place Stresses in the Single column: text box. Place 5 in the Subgroup size: text box. Place $\frac{\overline{R}}{d_n^*} = \frac{2.155}{2.326} = .92648$ in the Historical: sigma text box. Choose **OK**.

b. Calculate the appropriate control limits for an $\overline{X}$-chart based on R and plot the data. Comment on your results.

Follow these steps to construct an $\overline{X}$-chart based on the range.

Step 1. Construct the $\overline{X}$-chart.

Choose **Stat**>**Control Charts**>**Xbar-R**. Darken the Data are arranged as a Single column: option button. Place Stresses in the Single column: text box. Place 5 in the Subgroup size: text box. Place $\overline{X} = 2.63$ in the Historical mean: text box. Place $\frac{\overline{R}}{d_n^*} = \frac{2.155}{2.326} = .92648$ in the Historical: sigma text box. Choose **OK**.

Step 2. Try another approach.

Choose **Stat**>**Control Charts**>**Xbar-R**. Darken the Data are arranged as a Single column: option button. Place Stresses in the Single column: text box. Place 5 in the Subgroup size: text box. Clear the Historical mean: text box. Clear the Historical: sigma text box.

Use the sample pooled standard deviation.

Choose **Estimate**. Darken the Pooled standard deviation option button. Choose **OK**.

Choose **OK**.

c. What did you assume in order to construct these control limits? Given the information in the base period, how comfortable are you with these assumptions?

5.6 (EXO509) Snee (1983) examined the thicknesses of paint can ears. Periodically, the manufacturer took random samples of five cans each and measured the thickness of the ears. The data, in units of .001 inches, follow.

Sample	y_{i1}	y_{i2}	y_{i3}	y_{i4}	y_{i5}	Sample	y_{i1}	y_{i2}	y_{i3}	y_{i4}	y_{i5}
1	29	36	39	34	34	16	35	30	35	29	37
2	29	29	28	32	31	17	40	31	38	35	31
3	34	34	39	38	37	18	35	36	30	33	32
4	35	37	33	38	41	19	35	34	35	30	36
5	30	29	31	38	29	20	35	35	31	38	36
6	34	31	37	39	36	21	32	36	36	32	36
7	30	35	33	40	36	22	36	37	32	34	34
8	28	28	31	34	30	23	29	34	33	37	35
9	32	36	38	38	35	24	36	36	35	37	37
10	35	30	37	35	31	25	36	30	35	33	31
11	35	30	35	38	35	26	35	30	29	38	35
12	38	34	35	35	31	27	35	36	30	34	36
13	34	35	33	30	34	28	35	30	36	29	35
14	40	35	34	33	35	29	38	36	35	31	31
15	34	35	38	35	30	30	30	34	40	28	30

a. Calculate the appropriate control limits for an R-chart and plot the data. Comment on your results.

b. Calculate the appropriate control limits for an $\overline{X}$-chart based on R and plot the data. Comment on your results.

c. What did you assume in order to construct these control limits? Given the information in the base period, how comfortable are you with these assumptions?

5.7 (EXO508) A major manufacturer of writing instruments closely monitors the critical outside diameters for a popular pen barrel. This company uses an injection molding process to make these barrels as well as the caps. In order to guarantee the quality of the fit of the cap to the barrel, the company must keep the critical diameter of the barrel as consistent as possible. Each hour, the operator takes a random sample of three barrels and measures the critical outside diameter. The data for 25 such samples follow.

Sample	y_{i1}	y_{i2}	y_{i3}	Sample	y_{i1}	y_{i2}	y_{i3}
1	.379	.376	.379	14	.377	.380	.378
2	.378	.377	.378	15	.379	.378	.380
3	.378	.378	.378	16	.379	.381	.379
4	.378	.377	.377	17	.379	.381	.379
5	.378	.378	.378	18	.379	.378	.379
6	.378	.378	.377	19	.378	.379	.377
7	.379	.379	.379	20	.379	.379	.378
8	.379	.378	.377	21	.380	.378	.379
9	.378	.378	.377	22	.378	.381	.380
10	.377	.377	.378	23	.379	.380	.380
11	.381	.379	.377	24	.378	.379	.379
12	.379	.380	.379	25	.377	.377	.377
13	.378	.378	.379				

a. Calculate the appropriate control limits for an R-chart and plot the data. Comment on your results.

Follow these steps to construct an R-chart.

Step 1. Enter data.

Enter the data for each sample in column 1. Enter the first five observations for sample 1 followed by the five observations for sample 2, etc. Name the column *Diameter*.

Step 2. Construct the control chart.

Choose **Stat**>**Control Charts**>**R**. Darken the Data are arranged as a Single column: option button. Place Diameter in the Single column: text box. Place 3 in the Subgroup size: text box. Place $\frac{\bar{R}}{d_n^*} = \frac{0.00144}{1.693} = .00085$ in the Historical: sigma text box. Choose **OK**.

b. Calculate the appropriate control limits for an $\overline{X}$-chart based on R and plot the data. Comment on your results.

Follow these steps to construct an $\overline{X}$-chart based on the range.

Step 1. Construct the $\overline{X}$-chart.

Choose **Stat**>**Control Charts**>**Xbar-R**. Darken the Data are arranged as a Single column: option button. Place Diameter in the Single column: text box. Place 3 in the Subgroup size: text box. Place $\overline{X} = .37845$ in the Historical: mean: text box. Place $\frac{\bar{R}}{d_n^*} = \frac{0.00144}{1.693} = .00085$ in the Historical: sigma text box. Choose **OK**.

Step 2. Try another approach.

Choose **Stat**>**Control Charts**>**Xbar-R**. Darken the Data are arranged as a Single column: option button. Place Diameter in the Single column: text box. Place 3 in the Subgroup size: text box.

Use the sample pooled standard deviation.

Choose Estimate. Darken the Pooled standard deviation option button. Choose **OK**.

Choose **OK**.

c. What did you assume in order to construct these control limits? Given the information in the base period, how comfortable are you with these assumptions?

5.8 (EXO510) A small manufacturer of brake linings closely monitors the in- coming quality of a specific grade of graphite. This company receives the graphite in shipments of 20 pallets, with each pallet consisting of 40 fifty pound bags of graphite. An inspector takes a small amount of graphite from four randomly selected bags from each pallet and performs a standard ash test which burns off all the "volatiles" in the graphite. A low ash content indicates a high carbon content in the graphite. The following data represent the ash contents for the last shipment inspected.

Sample	y_{i1}	y_{i2}	y_{i3}	y_{i4}	y_{i5}	Sample	y_{i1}	y_{i2}	y_{i3}	y_{i4}	y_{i5}
1	19.2	19.5	19.3	19.3	19.2	11	18.8	19.2	18.6	18.7	18.8
2	19.0	18.7	18.9	18.3	19.0	12	19.9	20.1	20.4	20.0	19.9
3	19.0	18.5	18.4	18.6	19.0	13	19.2	19.2	19.0	19.1	19.2
4	19.1	19.0	19.0	18.9	19.1	14	20.6	20.1	20.0	20.2	20.6
5	18.5	18.4	18.3	18.4	18.5	15	20.2	19.9	19.7	19.7	20.2
6	18.9	18.7	18.7	18.6	18.9	16	20.0	19.6	19.6	19.6	20.0
7	19.8	19.4	19.3	19.3	19.8	17	19.9	19.8	19.7	19.8	19.9
8	19.3	19.5	19.2	19.2	19.3	18	20.1	19.8	19.8	19.7	20.1
9	19.6	19.6	20.2	19.3	19.6	19	20.0	19.9	19.9	20.6	20.0
10	18.8	19.2	19.1	18.8	18.8	20	20.1	20.0	19.8	19.9	20.1

a. Calculate the appropriate control limits for an R-chart and plot the data. Comment on your results.

Follow these steps to construct an R-chart.

Step 1. Enter data.

Enter the data for each sample in column 1. Enter the first five observations for sample 1 followed by the five observations for sample 2, etc. Name the column *AshCont*.

Step 2. Construct the control chart.

Choose **Stat**>**Control Charts**>**R**. Darken the Data are arranged as a Single column: option button. Place AshCont in the Single column: text box. Place 5 in the Subgroup size: text box. Place $\frac{\overline{R}}{d_n^*} = \frac{.440}{2.326} = .18917$ in the Historical: sigma text box.

Choose **OK**.

b. Calculate the appropriate control limits for an $\overline{X}$-chart based on R and plot the data. Comment on your results.

Follow these steps to construct an $\overline{X}$-chart based on the range.

Step 1. Construct the $\overline{X}$-chart.
Choose **Stat>Control Charts>Xbar-R**. Darken the Data are arranged as a Single column: option button. Place AshCont in the Single column: text box. Place 5 in the Subgroup size: text box. Place $\overline{X} = 19.41$ in the Historical mean: text box. Place $\dfrac{\overline{R}}{d_n^*} = \dfrac{.440}{2.326} = .18917$ in the Historical: sigma text box. Choose **OK**.

Step 2. Try another approach.
Choose **Stat>Control Charts>Xbar-R**. Darken the Data are arranged as a Single column: option button. Place AshCont in the Single column: text box. Place 5 in the Subgroup size: text box.
Use the sample pooled standard deviation.
Choose Estimate. Darken the Pooled standard deviation option button. Choose **OK**.
Choose **OK**.

3. What did you assume in order to construct these control limits? Given the information in the base period, how comfortable are you with these assumptions?

$\overline{X}-$ and s^2- Charts

While $\overline{X}$-charts and R-charts are used primarily out of tradition, a better approach might be to use $\overline{X}-$ and s^2- charts.

New Minitab Commands

1. **Stat>Control Charts>S** - Draws a control chart for subgroup standard deviations. In this section, you will use this command to produce this type of control chart in Example 5.3 - Grinding of Silicon Wafers.

2. **Stat>Control Charts>Xbar-S** - Draws a control chart for subgroup means (upper half of the screen) and a control chart for subgroup standard deviation (lower half of the screen) so you can examine both process level and process variation at the same time. In this section, you will use this command to produce this type of control chart in Example 5.3 - Grinding of Silicon Wafers.

Example 5.2 - Grinding of Silicon Wafers - Continued

From a statistical perspective the sample variance provides a far better measure of sample variability than the range. Minitab does not provide a s^2- chart but does provide a $s-$ chart.

Follow these steps to construct an S-chart.

Step 1. Use the data from the previous example.

Open the worksheet containing the data or enter the data for each batch in column 1. Enter the first five observations for batch 1 followed by the five observations for batch 2, etc. Name the column *Thickness*.

Step 2. Construct the control chart.

Choose **Stat**>**Control Charts**>**S**. Darken the Data are arranged as a Single column: option button. Place Thickness in the Single column: text box. Place 5 in the Subgroup size: text box. Choose Estimate. Darken the Pooled standard deviation option button. Choose **OK**. Choose **OK**.

The Minitab Output

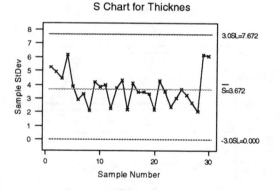

Figure 5.4

The Minitab output, as shown in Figure 5.4, indicates a process which appears to be in control. Observe that the UCL and the LCL are somewhat different than in the text. This is because Minitab uses a slightly different process to determine the upper and lower control limits. Consult the Minitab manual for more information in regards to the equations being used.

Minitab does produces a control chart for subgroup means and a control chart for the subgroup standard deviation.

Follow these steps to construct an Xbar-S chart.

Step 1. Use the data from the previous example.

Open the worksheet containing the data or enter the data for each batch in column 1. Enter the first five observations for batch 1 followed by the five observations for batch 2, etc. Name the column *Thickness*.

Step 2. Construct the control chart.

Choose **Stat**>**Control Charts**>**Xbar-S**. Darken the Data are arranged as a Single column: option button. Place Thickness in the Single column: text box. Place 5 in the Subgroup size: text box.

Use the sample pooled standard deviation.

Choose **Estimate**. Darken the **P**ooled standard deviation option button.

Choose **OK**.

Choose **OK**.

The Minitab Output

Xbar/S Chart for Thicknes

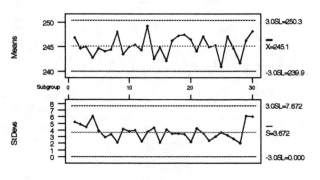

Figure 5.5

The Minitab output, as shown in Figure 5.5, again indicates a process which appears to be in control.

Exercises

5.9 (EXO5O7) Padgett and Spurrier (1990) analyze the breaking strengths of carbon fibers used in fibrous composite materials. These fibers measure 50 mm in length and 7-8 microns in diameter. Periodically, the manufacturer selects random samples of five fibers and tests their breaking stresses. Specifications require that 99 of the fibers must have a breaking stress of at least 1.2 GPa (giga-Pascals). The breaking stresses in GPa from 20 such samples follow.

Sample	y_{i1} y_{i2} y_{i3} y_{i4} y_{i5}	Sample	y_{i1} y_{i2} y_{i3} y_{i4} y_{i5}
1	3.7 2.7 2.7 2.5 3.6	11	1.4 3.7 3.0 1.4 1.0
2	3.1 3.3 2.9 1.5 3.1	12	2.8 4.9 3.7 1.8 1.6
3	4.4 2.4 3.2 3.2 1.7	13	3.2 1.6 0.8 5.6 1.7
4	3.3 3.1 1.8 3.2 4.9	14	1.6 2.0 1.2 1.1 1.7
5	3.8 2.4 3.0 3.0 3.4	15	2.2 1.2 5.1 2.5 1.2
6	3.0 2.5 2.7 2.9 3.2	16	3.5 2.2 1.7 1.3 4.4
7	3.4 2.8 4.2 3.3 2.6	17	1.8 0.4 3.7 2.5 0.9
8	3.3 3.3 2.9 2.6 3.6	18	1.6 2.8 4.7 2.0 1.8
9	3.2 2.4 2.6 2.6 2.4	19	1.6 1.1 2.0 1.6 2.1
10	2.8 2.8 2.2 2.8 1.9	20	1.9 2.9 2.8 2.1 3.7

a. Calculate the appropriate control limits for an s-chart and plot the data. Comment on your results.

Follow these steps to construct an S-chart.

Step 1. Enter data.

Enter the data for each batch in column 1. Enter the first five observations for batch 1 followed by the five observations for batch 2, etc. Name the column *Stress*.

Step 2. Construct the control chart.

Choose **Stat**>**Control Charts**>**S**. Darken the Data are arranged as a Single column: option button. Place Stress in the Single column: text box. Place 5 in the Subgroup size: text box.

Use the sample pooled standard deviation.

Choose Estimate. Darken the Pooled standard deviation option button. Choose **OK**.

Choose **OK**.

b. Calculate the appropriate control limits for an $\overline{X}$-chart based on s and plot the data. Comment on your results.

Follow this step to construct an Xbar-S chart.

Choose **Stat**>**Control Charts**>**Xbar-S**. Darken the Data are arranged as a Single column: option button. Place Stress in the Single column: text box. Place 5 in the Subgroup size: text box.

Use the sample pooled standard deviation.

Choose Estimate. Darken the Pooled standard deviation option button. Choose **OK**.

Choose **OK**.

c. What did you assume in order to construct these control limits? Given the information in the base period, how comfortable are you with these assumptions?

5.10 (EXO5O9) Snee (1983) examined the thicknesses of paint can ears. Periodically, the manufacturer took random samples of five cans each and measured the thickness of the ears. The data, in units of .001 inches, follow.

Sample	y_{i1}	y_{i2}	y_{i3}	y_{i4}	y_{i5}	Sample	y_{i1}	y_{i2}	y_{i3}	y_{i4}	y_{i5}
1	29	36	39	34	34	16	35	30	35	29	37
2	29	29	28	32	31	17	40	31	38	35	31
3	34	34	39	38	37	18	35	36	30	33	32
4	35	37	33	38	41	19	35	34	35	30	36
5	30	29	31	38	29	20	35	35	31	38	36
6	34	31	37	39	36	21	32	36	36	32	36
7	30	35	33	40	36	22	36	37	32	34	34
8	28	28	31	34	30	23	29	34	33	37	35
9	32	36	38	38	35	24	36	36	35	37	37
10	35	30	37	35	31	25	36	30	35	33	31
11	35	30	35	38	35	26	35	30	29	38	35
12	38	34	35	35	31	27	35	36	30	34	36
13	34	35	33	30	34	28	35	30	36	29	35
14	40	35	34	33	35	29	38	36	35	31	31
15	34	35	38	35	30	30	30	34	40	28	30

a. Calculate the appropriate control limits for an R-chart and plot the data. Comment on your results.

b. Calculate the appropriate control limits for an $\overline{X}$-chart based on R and plot the data. Comment on your results.

c. What did you assume in order to construct these control limits? Given the information in the base period, how comfortable are you with these assumptions?

5.11 (EXO5O8) A major manufacturer of writing instruments closely monitors the critical outside diameters for a popular pen barrel. This company uses an injection molding process to make these barrels as well as the caps. In order to guarantee the quality of the fit of the cap to the barrel, the company must keep the critical diameter of the barrel as consistent as possible. Each hour, the operator takes a random sample of three barrels and measures the critical outside diameter. The data for 25 such samples follow.

Sample	y_{i1} y_{i2} y_{i3}	Sample	y_{i1} y_{i2} y_{i3}
1	.379 .376 .379	14	.377 .380 .378
2	.378 .377 .378	15	.379 .378 .380
3	.378 .378 .378	16	.379 .381 .379
4	.378 .377 .377	17	.379 .381 .379
5	.378 .378 .378	18	.379 .378 .379
6	.378 .378 .377	19	.378 .379 .377
7	.379 .379 .379	20	.379 .379 .378
8	.379 .378 .377	21	.380 .378 .379
9	.378 .378 .377	22	.378 .381 .380
10	.377 .377 .378	23	.379 .380 .380
11	.381 .379 .377	24	.378 .379 .379
12	.379 .380 .379	25	.377 .377 .377
13	.378 .378 .379		

a. Calculate the appropriate control limits for an s-chart and plot the data. Comment on your results.

b. Calculate the appropriate control limits for an $\overline{X}$-chart based on s and plot the data. Comment on your results.

c. What did you assume in order to construct these control limits? Given the information in the base period, how comfortable are you with these assumptions?

5.12 (EXO510) A small manufacturer of brake linings closely monitors the in- coming quality of a specific grade of graphite. This company receives the graphite in shipments of 20 pallets, with each pallet consisting of 40 fifty pound bags of graphite. An inspector takes a small amount of graphite from four randomly selected bags from each pallet and performs a standard ash test which burns off all the "volatiles" in the graphite. A low ash content indicates a high carbon content in the graphite. The following data represent the ash contents for the last shipment inspected.

Sample	y_{i1} y_{i2} y_{i3} y_{i4} y_{i5}	Sample	y_{i1} y_{i2} y_{i3} y_{i4} y_{i5}
1	19.2 19.5 19.3 19.3 19.2	11	18.8 19.2 18.6 18.7 18.8
2	19.0 18.7 18.9 18.3 19.0	12	19.9 20.1 20.4 20.0 19.9
3	19.0 18.5 18.4 18.6 19.0	13	19.2 19.2 19.0 19.1 19.2
4	19.1 19.0 19.0 18.9 19.1	14	20.6 20.1 20.0 20.2 20.6
5	18.5 18.4 18.3 18.4 18.5	15	20.2 19.9 19.7 19.7 20.2
6	18.9 18.7 18.7 18.6 18.9	16	20.0 19.6 19.6 19.6 20.0
7	19.8 19.4 19.3 19.3 19.8	17	19.9 19.8 19.7 19.8 19.9
8	19.3 19.5 19.2 19.2 19.3	18	20.1 19.8 19.8 19.7 20.1
9	19.6 19.6 20.2 19.3 19.6	19	20.0 19.9 19.9 20.6 20.0
10	18.8 19.2 19.1 18.8 18.8	20	20.1 20.0 19.8 19.9 20.1

a. Calculate the appropriate control limits for an s-chart and plot the data. Comment on your results.

Follow these steps to construct an S-chart.

Step 1. Enter data.

Enter the data for each sample in column 1. Enter the first five observations for sample 1 followed by the five observations for sample 2, etc. Name the column *AshCont*.

Step 2. Construct the control chart.

Choose **Stat**>**Control Charts**>**S**. Darken the Data are arranged as a Single column: option button. Place AshCont in the Single column: text box. Place 5 in the Subgroup size: text box.

Use the sample pooled standard deviation.

Choose **Estimate**. Darken the Pooled standard deviation option button. Choose **OK**.

Choose **OK**.

b. Calculate the appropriate control limits for an $\overline{X}$-chart based on s and plot the data. Comment on your results.

Follow this step to construct an Xbar-S chart.

Choose **Stat**>**Control Charts**>**Xbar-S**. Darken the Data are arranged as a Single column: option button. Place Stress in the Single column: text box. Place 5 in the Subgroup size: text box.

Use the sample pooled standard deviation.

Choose **Estimate**. Darken the Pooled standard deviation option button. Choose **OK**.

Choose **OK**.

c. What did you assume in order to construct these control limits? Given the information in the base period, how comfortable are you with these assumptions?

X-Chart

Some processes do not produce data frequently enough to justify monitoring sample means. In such cases, each individual observation is used to construct a control chart.

New Minitab Commands

1. **Stat**>**Control Charts**>**Individuals** - Draws a control chart for individual observations. In this section, you will produce this type of control chart in Example 5.4 - Viscosities from a Batch Chemical Process.

Example 5.4 - Viscosities from a Batch Chemical Process

Holmes and Mergen (1992) studied a batch operation at a chemical plant where an important quality characteristic was the product viscosity. At the end of each

12 hour batch, an operator took a viscosity measurement. Since data from this process come infrequently, management required a monitoring procedure based on the individual viscosities. The 140 observations are listed in order column by column.

13.3	14.5	15.3	15.3	14.3	14.8	15.2
14.9	14.6	14.1	14.3	16.1	13.1	15.5
12.6	14.6	14.3	15.4	15.2	16.8	14.9
13.7	15.2	14.5	15.3	15.6	15.8	13.3
14.1	15.4	15.2	15.2	15.9	16.5	14.0
15.1	17.0	14.9	14.8	14.0	15.8	13.7
15.1	13.4	14.1	14.8	14.3	14.3	16.4
16.9	14.2	16.9	14.9	15.2	14.4	15.2
14.6	16.4	14.2	15.7	16.0	14.9	13.6
15.3	14.3	15.6	16.1	13.9	15.2	14.4
14.0	14.4	13.7	13.8	15.6	14.5	12.8
16.1	16.6	15.6	15.7	13.0	13.9	16.2
13.8	16.5	14.2	14.9	14.7	15.0	14.4
14.4	15.4	16.3	15.0	15.7	14.5	15.5
14.4	14.4	14.8	15.6	14.5	14.9	16.0
15.0	14.7	15.1	15.4	16.0	18.6	16.0
15.9	14.5	15.1	14.2	17.6	13.5	15.3
15.0	15.6	15.7	16.4	14.5	14.9	14.6
15.5	14.7	15.0	13.8	14.0	15.8	14.8
15.8	16.7	16.4	15.3	15.7	15.0	16.8

Follow these steps to construct an X-chart.

Step 1. Enter data.

Enter each observation, proceeding vertically down each column, and then moving from column to column, in column C1. Name column C1 as *Viscoty*.

Step 2. Construct the X-chart.

Choose **Stat**>**Control Charts**>**Individuals**. Place Viscoty in the Variable: text box. Place 14.87 in the Historical mean: text box.

Use a moving range.

Choose **Estimate**. Place a check in the Use moving range of length: check box. Place 40 in the Use moving range of length: text box. Choose **OK**. Choose **OK**.

The Minitab Output

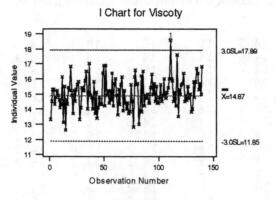

Figure 5.6

The Minitab output, as shown in Figure 5.6, illustrates a control chart for all 140 observations suggesting that sample 111 is a possible out of control situation which requires attention. The chart suggests that the viscosity for this batch has drifted higher than normal.

Exercises

5.13 (EXO519) Yashchin (1992) monitored the thicknesses of metal wires produced in a chip-manufacturing process. Ideally, these wires should have a target thickness of 8 microns. The data, in microns, follow. The data are in consecutive order, reading across the rows. The first observation is 8.4, the second is 8.0, etc. Use the first 30 observations as a base period.

8.4 8.0 7.8 8.0 7.9 7.7 8.0 7.9 8.2 7.9

7.9 8.2 7.9 7.8 7.9 7.9 8.0 8.0 7.6 8.2

8.1 8.1 8.0 8.0 8.3 7.8 8.2 8.3 8.0 8.0

7.8 7.9 8.4 7.7 8.0 7.9 8.0 7.7 7.7 7.8

7.8 8.2 7.7 8.3 7.8 8.3 7.8 8.0 8.2 7.8

Use the first 30 observations as a base period.

    a. Calculate the appropriate control limits and plot the X-chart. Comment on your results.

    Follow these steps to construct an X-chart.
      Step 1. Enter data.
            Enter the data in column 1. Enter the first observation, 8.4, followed by the second observation, 8.0, etc. Name the column *Thickness*.

Step 2. Construct the X-chart.

Choose **Stat**>**Control Charts**>**Individuals**. Place Thickness in the Variable: text box. Place 8 in the Historical mean: text box.

Use a moving range.

Choose **Estimate**. Place a check in the Use moving range of length: check box. Place 30 in the Use moving range of length: text box. Choose **OK**.

Choose **OK**.

b. What did you assume in order to construct these control limits? Given the information in the base period, how comfortable are you with these assumptions?

5.14 (EXO520) Cryer and Ryan (1990) monitor a chemical process where the quality characteristic is a color property. The data are in consecutive order, reading across the rows. The first observation is 0.67, the second is 0.63, etc.

0.67 0.63 0.76 0.66 0.69 0.71 0.72
0.71 0.72 0.72 0.83 0.87 0.76 0.79
0.74 0.81 0.76 0.77 0.68 0.68 0.74
0.68 0.69 0.75 0.80 0.81 0.86 0.86
0.79 0.78 0.77 0.77 0.80 0.76 0.67

Use the first 25 observations as a base period.

a. Calculate the appropriate control limits and plot the X-chart. Comment on your results.

b. What did you assume in order to construct these control limits? Given the information in the base period, how comfortable are you with these assumptions?

5.15 (EXO521) Van Nuland (1992) daily compares two temperature instruments: one coupled to a process computer, and the other used for visual control. Ideally, these two instruments should agree. As a result, he monitors the daily difference in the temperature readings (he is using a control chart based on **paired differences**). The data for 35 days follow. The data are in consecutive order, reading across the rows. The first observation is 0.3, the second observation is 0.0, etc.

0.3 0.0 0.1 0.3 0.3
0.5 0.2 0.1 0.3 0.0
-0.1 0.5 0.4 0.1 0.1
-0.1 0.4 0.1 0.2 0.0
0.1 0.3 0.2 0.1 0.4
0.2 0.4 0.0 0.2 0.4
0.6 0.6 0.5 0.7 0.7

Use the first 30 observations as a base period.

a. Calculate the appropriate control limits and plot the X-chart. Comment on your results.

Follow these steps to construct an X-chart.

Step 1. Enter data.
Enter the data in column 1. Enter the first observation, 0.3, followed by the second observation, 0.0, etc. Name the column *Diff*.

Step 2. Construct the X-chart.
Choose **Stat**>**Control Charts**>**Individuals**. Place Diff in the Variable: text box. Place 0 in the Historical mean: text box.
Use a moving range.
Choose **Estimate**. Place a check in the Use moving range of length: check box. Place 35 in the Use moving range of length: text box. Choose **OK**.
Choose **OK**.

b. What did you assume in order to construct these control limits? Given the information in the base period, how comfortable are you with these assumptions?

5.16 (EXO522) King (1992) monitors the net weights of a nominally 16 oz. packaged product. An inspector collected a sample of 20 packages and accurately measured their net contents. The data are in consecutive order, reading across the rows. The first observation is 16.4, the second observation is 16.4, etc.

16.4 16.4 16.5 16.5 16.6 16.7 16.2 16.4 16.4 16.5
16.6 16.6 16.8 16.3 16.4 16.5 16.5 16.6 16.7 16.8

Use all the data as a base period.

a. Calculate the appropriate control limits and plot the X-chart. Comment on your results.

Follow these steps to construct an X-chart.

Step 1. Enter data.
Enter the data in column 1. Enter the first observation, 16.4, followed by the second observation, 16.4, etc. Name the column *Weights*.

Step 2. Construct the X-chart.
Choose **Stat**>**Control Charts**>**Individuals**. Place Weights in the Variable: text box. Place 16 in the Historical mean: text box.
Use a moving range.

Choose Estimate. Place a check in the Use moving range of length: check box. Place 20 in the Use moving range of length: text box. Choose **OK**.
Choose **OK**.

b. What did you assume in order to construct these control limits? Given the information in the base period, how comfortable are you with these assumptions?

5.3   Attribute Charts

New Minitab Commands

1. **Stat>Control Charts>NP** - Draws a chart for the number of defectives, where the sample size is the same each time. In this section, you will use this command in Example 5.4 -Nonconforming Bricks.

2. **Stat>Control Charts>P** - Draws a chart for the proportion of defectives, where the sample size is not the same each time.

The np-Chart

We can develop a control chart to monitor the number of items which fail to meet specifications, which may be well-modeled by a binomial distribution.

Example 5.4 - Non-Conforming Brick

Marcucci (1985) reports on a brick manufacturing process which classifies the product as either

a. conforming (suitable for all purposes),
b. non-conforming (structurally sound but not suitable for all uses, or unacceptable for use (cull).

Sixteen samples, each of size 200 are reported with the following number of non-conforming bricks in each sample.

Sample	y_i	Sample	y_i
1	9	9	13
2	8	10	31
3	12	11	18
4	8	12	15
5	16	13	15
6	9	14	16
7	11	15	10
8	12	16	9

Follow these steps to construct a np-chart.

Step 1. Enter data.

Enter the data for each sample in column 1. Name column C1 as *Defects*.

Step 2. Construct the control chart.

Choose **Stat**>**Control Charts**>**NP**. Place Defects in the Variable: text box. Darken the S̲ubgroup si̲ze: option button. Place 200 in the S̲ubgroup si̲ze: text box.

Select the subgroup size.

Choose E̲stimate. Darken the S̲ubgroup size: option button. Place 16 in the S̲ubgroup size: text box. Choose **OK**.

Choose **OK**.

The Minitab Output

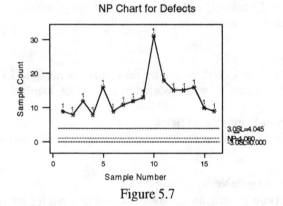

Figure 5.7

The Minitab output, as shown in Figure 5.7, indicates clearly that sample 10 is a possible out of control situation which requires attention. The chart suggests that the number of non-conforming bricks from that day has drifted to a higher than normal level.

Exercises

5.17 (EXO523) Automobile hub caps typically are produced by a metal casting process which historically has been slow and expensive. A common problem facing many older casting processes is "flashing." In casting, liquid metal is shot into a mold and rapidly cooled. A flash commonly forms on the piece at the spot in the mold where the metal flows. A major automobile manufacturer closely monitors the incoming quality of the hub caps coming from a particular supplier by inspecting 200 hub caps from every incoming shipment. The following data are the number of hub caps with at least minor flashing for the last 40 shipments. The data are in consecutive order, reading across the rows.

The first observation is 20, the second is 23, etc.

20 23 20 15 25 24 27 18 17 20
26 15 20 21 15 18 12 25 16 25
24 27 21 19 14 23 17 20 19 20
23 18 25 22 17 20 22 24 15 11

Use the first 30 samples as a base period. Calculate the appropriate control limits and plot the control chart. Comment on your results.
Follow these steps to construct a np-chart.

Step 1. Enter data.

Enter the data for each sample in column 1. Name column C1 as *Flash*.

Step 2. Construct the control chart.

Choose **Stat**>**Control Charts**>**NP**. Place Flash in the Variable: text box. Darken the Subgroup size: option button. Place 200 in the Subgroup size: text box.
Select the subgroup size.
Choose Estimate. Darken the Subgroup size: option button. Place 40 in the Subgroup size: text box. Choose **OK**.
Choose **OK**.

5.18 (EXO524) A manufacturer of nickel-hydrogen batteries discovered a problem with "blisters" on its nickel plates. These blisters would cause the resulting battery cell to short out prematurely. Each week, the manufacturer randomly selects 100 plates, constructs test cells, cycles these cells 50 times, and counts the number of plates which blister. The following data are the number of plates which blister for a 26 week period. The data are in consecutive order, reading across the rows. The first observation is 5, the second is 15, etc.

5   15   7   11   3   12   7   11   12
7   12   8   10   8   6   4   5   7
9   9   8   0 11   11   8   7

. Use the first 20 samples as a base period. Calculate the appropriate control limits and plot the control chart. Comment on your results.

5.19 (EXO525) Felt-tip markers have shelf lives of approximately two years. Most manufacturers used accelerated life testing whereby markers are placed in an oven at elevated temperatures for a given period of time, usually on the order of six weeks. The proportion which survive the elevated temperatures provides a good estimate of the proportion which should survive two years on a shelf. One major writing instrument company performs an accelerated life test on a random sample of 300 markers from each lot of markers. The following data are the number of markers which fail the accelerated life test for the last 40 lots.

The data are in consecutive order, reading across the rows. The first observation is 14, the second is 16, etc.

14	16	9	14	17	13	14	19	16	11
8	11	17	5	19	17	18	17	22	18
12	16	15	12	15	16	14	20	20	17
15	14	19	13	19	19	23	18	18	21

Use the first 30 samples as a base period. Calculate the appropriate control limits and plot the control chart. Comment on your results.

5.20 (EXO526) A major manufacturer of writing instruments uses high speed equipment to assemble pencils. Each day, the operator randomly selects 10 gross of pencils (10*144 or 1440) pencils and classifies each pencil as either OK or non-conforming. The following data are the number of non-conforming pencils for each of 30 days of production. The data are in consecutive order, reading across the rows. The first observation is 17, the second is 9, etc.

17	9	15	17	12	12	17	15	15	10
14	9	15	10	15	17	21	14	18	10
19	21	20	26	19	21	18	20	15	20

Use the first 20 samples as a base period. Calculate the appropriate control limits and plot the control chart. Comment on your results.

5.21 (EXO529) In some cases, we cannot get the same size sample each time. The proper control chart must adapt the control limits for the actual sample size used (called a **p-chart**. For example, the actual numbers of non-conforming brick from Marcucci (1985) came from samples of differing size. Let n_i and let P_i be the actual sample size and the proportion of non-conforming brick, respectively, for the ith sample. Use the hypothesis test for proportions outlined in Section 6.4 of the text to develop an appropriate monitoring procedure based on the p_i's. Apply this procedure to the actual Marcucci (1985) data, which follow.

Sample	n_i	y_i	Sample	n_i	y_i
1	254	12	9	221	14
2	207	8	10	206	32
3	243	15	11	245	22
4	201	8	12	221	17
5	232	18	13	212	16
6	138	6	14	245	20
7	218	12	15	237	12
8	155	9	16	148	7

Follow these steps to construct a *p*-chart using Minitab.

Step 1. Enter data.
Enter the data for each sample in columns 1, 2 and 3. Name column C1 as Sample, column C2 as Size and column C3 as *Defects*.

Step 2. Construct the control chart.
Choose **Stat**>**Control Charts**>**P**. Place Defects in the Variable: text box. Darken the S**u**bgroups in: option button. Place Size in the S**u**bgroups in: text box.
Select the subgroup size.
Choose **E**stimate. Darken the A**c**tual subgroup sizes option button.
Choose **OK**.
Choose **OK**.

The *c*-Chart

New Minitab Commands

1. **Stat**>**Control Charts**>**C** - Draws a chart of the number of defects. The column containing the number of defectives for one sample is assumed to have come from a Poisson distribution with parameter μ. In this section, you will use this command to produce a control chart in Example 5.4 - Industrial Accident Data. We can develop a control chart to monitor small counts such as the number of incidents per period or the number of non-conformances per unit, where the size of the period or unit remains constant. These incidents can often be well-modeled by a Poisson distribution.

Example 5.5 - Industrial Accident Data

Lusas (1985) studied the number of accidents for a ten year period at a DuPont facility. DuPont historically has strongly emphasized the importance of safety and closely monitor the accident rate. The following table gives the number of accidents each calender quarter (three month period).

Quarter	Accidents	Quarter	Accidents	Quarter	Accidents	Quarter	Accidents
1	5	11	6	21	3	31	1
2	5	12	9	22	4	32	4
3	10	13	5	23	2	33	1
4	8	14	6	24	0	34	2
5	4	15	5	25	1	35	2
6	5	16	10	26	3	36	1
7	7	17	6	27	2	37	4
8	3	18	3	28	2	38	4
9	2	19	3	29	7	39	4
10	8	20	10	30	7	40	4

Follow these steps to construct a c-chart.

Step 1. Enter data.

Enter the data for each Quarter in column 1. Name column C1 as Accident.

Step 2. Construct the control chart.

Choose **Stat**>**Control Charts**>**C**. Place Accident in the Variable: text box.

Consider the first 20 calendar quarters as the base period.

Choose **Estimate**. Place 21:40 in the O**m**it the following samples when estimating parameters: text box. Choose **OK**.

Choose **OK**.

The Minitab Output

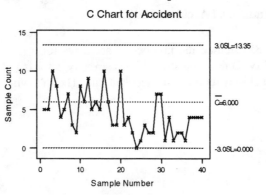

Figure 5.8

The Minitab output, as shown in Figure 5.8, indicates a process that looks well in control over the base period. However, after the base period, only two observations are larger than $\bar{c}$, the average count over the base period. The other observations all fall between $\bar{c}$ and the lower control limit, which is a clear sign that the accident rate has dropped at this facility. Such a reduction is a clear indication that the facilities efforts to improve safety are working.

Exercises

5.22 (EXO531) The manufacture of silicon wafers used in integrated circuits requires the removal of contaminating particles of a certain size. Yashchin (1995)
monitored a rinsing process for these wafers. This process rinses batches of 20
wafers with deionized water. The process then dries these wafers by spinning
off the water droplets. Prior to loading the wafers in the rinser/dryer, production
personnel count the number of contaminating particles. This count provides
feedback on the cleanliness of the manufacturing environment. The following
data represent the counts per batch for 60 successive batches. The data are
in consecutive order, reading across the rows. The first observation is 7, the
second is 4, etc.

7	4	9	9	2	10	3	6	6	5
5	7	5	7	3	4	8	4	5	5
9	8	8	8	13	10	6	10	11	3
11	13	9	11	13	15	6	10	11	12
12	2	4	7	2	4	7	6	7	4
6	4	6	6	8	5	6	9	3	6

Use the first 20 samples as a base period. Calculate the appropriate control
limits and plot the control chart. Comment on your results.
Follow these steps to construct a c-chart.

   Step 1. Enter data.
           Enter the data for each batch in column 1. Name column C1 as *Par-
           ticle*.

   Step 2. Construct the control chart.
           Choose **Stat**>**Control Charts**>**C**. Place Particle in the Variable: text
           box.
           Use the first 20 samples as a base period.
           Choose Estimate. Place 21:40 in the Omit the following samples
           when estimating parameters: text box. Choose **OK**.
           Choose **OK**.

5.23 (EXO530) Nelson (1987) considers a process where an important quality characteristic is the number of flaws per length of wire. Routinely, inspectors examine 5000-meter lengths of wire and count the number of flaws found. The
following data are the number of flaws for the last 30 sections inspected. The
data are in consecutive order, reading across the rows. The first observation is
15, the second is 7, etc.

15	7	13	13	5	8	15	10	10	7
14	16	15	14	21	10	15	15	13	24
22	18	18	14	8	11	6	10	1	3

Use the first 20 samples as a base period. Calculate the appropriate control limits and plot the control chart. Comment on your results.

5.24 (EXO532) A major automobile manufacturer inspects one car an hour for minor defects as the car rolls off the final assembly line. Virtually all of these defects are minor, usually cosmetic. The following data are the number of defects found on the last 40 cars inspected. The data are in consecutive order, reading across the rows. The first observation is 4, the second is 7, etc.

4	7	5	6	9	5	10	5	6	9
7	5	6	2	4	8	9	7	7	6
8	9	8	6	7	9	5	8	6	15
1	6	5	6	6	4	10	3	7	3

Use the first 30 samples as a base period. Calculate the appropriate control limits and plot the control chart. Comment on your results.

Chapter 6
Linear Regression Analysis

6.1   Overview

Relationships Among Data

Engineers use models, which express the relationships among various characteristics, to predict a characteristic of interest, called the response or dependent variable. The characteristics used to predict the response variable are called the independent or predictor variables. Minitab can assist engineers to

1. build models;
2. examine the relationship between the response and predictor variables;
3. determine the adequacy of the model;
4. make predictons.

Minitab will perform simple linear correlation(s), linear regression and multiple regression. Both numerical and graphical presentations are available. After reading this chapter, you should be able to

- Construct a scatter plot of data.
- Determine the simple linear regression equation for bivariate data.
- Construct a fitted line plot.
- Assess the adequacy of the model.
- Determine confidence intervals and prediction intervals.
- Determine the least squares multiple regression equation for more than one independent variable.
- Perform a residual analysis of the data.
- Transform data.

6.2   Overview of Simple Linear Regression

New Minitab Commands

1. **Stat**>**Regression**>**Regression** - Performs simple, polynomial regression, and multiple regression using the least squares method. In this section, you will use this command to obtain the regression equation for the data in Example 6.1 -Vapor Pressure of Water.

2. **Stat**>**Regression**>**Fitted Line Plot** - Fits a simple linear or polynomial (second or third order) regression model and plots a regression line through the actual data or the log10 of the data. The fitted line plot shows you how closely

the actual data lie to the fitted regression line. In this section, you will obtain a fitted line plot to illustrate how the estimated relationship fits the data in a simple linear regression model.

Scatter Plots

A scatterplot is a graph of the relationship between the two characteristics of interest. The scatterplot provides a visual means of assessing the relationship between the variables and can assist us in proposing reasonable models.

Example 6.1 - Vapor Pressure of Water

The vapor pressure of water for specific temperatures is often required in many situtations. The following table summarizes the vapor pressure of water for various temperatures from 0^0C to 60^0C.

Temp. Deg. C	Vapor Pres. mm Hg
10	9.2
20	17.5
30	31.8
40	55.3
50	92.5
60	149.4

Follow these steps to construct a scatter plot of the data.

Step 1. Enter data.

Enter the temperatures in column C1. Name column C1 as *Temp*. Enter the vapor pressures in column C2. Name column C2 as *Pressure*.

Step 2. Construct the scatter plot.

Choose **Graph**>**Plot**. Place Pressure in the Y **G**raph variables: text box and Temp in the X **G**raph variables: text box.

Enter a title.

Choose **A**nnotation>**T**itle. Type the first title of *Vapor Pressure of Water* in the Title text box. Choose **OK**.

Choose **OK**.

The Minitab Output

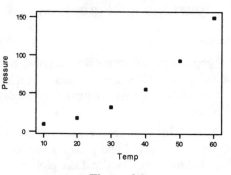

Figure 6.1

The scatter plot, as shown in Figure 6.1, indicates that as temperature increases, so does the vapor pressure. The plot suggests that it may be possible to model the relationship between vapor pressure and temperature using a straight line as a first approximation.

Simple Linear Regression Model

We can model the response as a linear relationship of the independent variable. The simple linear regression model is a straight line of the form

$$y_i = \beta_0 + \beta_1 x_i + \epsilon_u$$

where

1. β_0 is the y-intercept,
2. β_1 is the slope, and
3. ϵ_i is a random error.

Follow this step to determine the least squares equation between vapor pressure and temperature.

Choose **Stat**>**Regression**>**Regression**. Place Pressure in the Response: text box. Place Temp in the Predictors: text box. Choose **OK**.

The Minitab Output

Regression Analysis

```
The regression equation is
Pressure = - 35.7 + 2.71 Temp

Predictor        Coef        StDev           T          P
Constant       -35.67        17.23       -2.07      0.107
Temp           2.7129       0.4425        6.13      0.004

S = 18.51       R-Sq = 90.4%      R-Sq(adj) = 88.0%

Analysis of Variance

Source        DF          SS          MS          F          P
Regression     1       12879       12879      37.59      0.004
Error          4        1370         343
Total          5       14250
```

Figure 6.2

The Minitab output, as shown in Figure 6.2, indicates the regression equation between pressure ($\widehat{y}$) and temperature (x). The prediction equation might be written as

$$\widehat{y} = -35.7 + 2.71x.$$

A plot illustrating how the estimated relationship fits the data is possible in Minitab. This plot is called a fitted line plot.

Follow this step to construct the fitted line plot.
Choose **Stat**>**Regression**>**Fitted Line Plot.** Place Pressure in the Response (Y): text box. Place Temp in the Predictor (X): text box. Darken the Linear Type of Regression Model option button.
Enter a title.
Choose Options. Place *Vapor Pressure of Water* in the Title: text box. Choose **OK**
Choose **OK.**

The Minitab Output

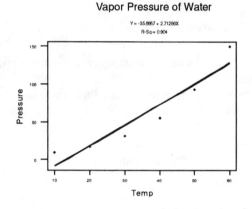

Figure 6.3

The Minitab output, as shown in Figure 6.3, indicates that the straight line does not perfectly fit the vapor pressure data. Rather, the plot indicates that the true relationship between vapor pressure and temperature is somewhat non-linear. The plot suggests that additional models should be explored.

6.3   Analysis of the Model

You have now worked with Minitab to perform simple linear regression problems that relate one variable to another variable. Now it is appropriate to begin to assess the adequacy of the model. Specifically, we will perform hypothesis tests for the slope of the regression line.

Example 6.2 - Vapor Pressure of Water - - Testing the Slope
A reasonable question for the vapor pressure - temperature data considers the specific nature of the relationship between the variables. Specifically, the null hypothesis is that vapor pressure does not change as a function of temperature, i.e., $H_0 : \beta_1 = 0$. The steps to obtain the least squares regression equation between vapor pressure and temperature resulted in the following Minitab output.

The Minitab Output

Regression Analysis

```
The regression equation is
Pressure = - 35.7 + 2.71 Temp
```

Predictor	Coef	StDev	T	P
Constant	-35.67	17.23	-2.07	0.107
Temp	2.7129	0.4425	6.13	0.004

```
S = 18.51      R-Sq = 90.4%      R-Sq(adj) = 88.0%
```

Analysis of Variance

Source	DF	SS	MS	F	P
Regression	1	12879	12879	37.59	0.004
Error	4	1370	343		
Total	5	14250			

Figure 6.4

The Minitab output, as shown in Figure 6.4, indicates that the test statistic, t, for the slope is 6.13, with a p-value of 0.004. The small p-value would indicate that the null hypothesis is rejected and we may conclude that vapor pressure does increase as temperature increases.

The Coefficient of Determination and the Overall F Test
Vapor Pressure of Water - R^2

In any assessment of the adequacy of the model, it is most appropriate to examine the analysis of variance table for regression as well as the coefficient of determination.

The coefficient of determination, R^2, is an indication of the proportion of the total variablity explained by the regression model. For the analysis of vapor pressure and temperature the previous Minitab output indicates that the coefficient of determination is 90.4%. Thus, the linear model explains 90.4% of the total variation.

The analysis of variance table for the regression of pressure on temperature is indicative of the overall adequacy of the model. Once the least squares line for the data set has been determined, the general approach is to decide whether or not this line describes a statistically significant linear relationship. If the assumption is that $y_i = \beta_0 + \beta_1 x_i + \epsilon_u$, then deciding whether or ot a statistically significant linear relationship exists between the variables is equivalent to deciding whether or not $\beta_1 = 0$. For the analysis of vapor pressure and temperature the previous Minitab output indicates that the F statistic is 37.59 with the associated p-value of 0.004. Thus, the regression of pressure on temperature is significant.

Confidence and Prediction Intervals

Once the equation for the line of best fit has been obtained and determined to be usable, the equation may be used to make predictions. A confidence interval estimates the mean of the population of y values at a given value of x, while a prediction interval estimates the individual y value for a given value of x.

Follow this step to determine confidence intervals and prediction intervals.

Obtain the regression equation.
Choose **Stat**>**Regression**>**Regression.** Place Pressure in the Response: text box. Place Temp in the Predictors: text box.
Set up the prediction and confidence intervals.
Choose Options. Place Temp in the Prediction intervals for new observations: text box. Choose the default value of 95 (%) in the Confidence level: text box. Choose **OK.** Choose **OK.**

The Minitab Output

Regression Analysis

The regression equation is
Pressure = - 35.7 + 2.71 Temp

Predictor	Coef	StDev	T	P
Constant	-35.67	17.23	-2.07	0.107
Temp	2.7129	0.4425	6.13	0.004

S = 18.51   R-Sq = 90.4%   R-Sq(adj) = 88.0%

Analysis of Variance

Source	DF	SS	MS	F	P
Regression	1	12879	12879	37.59	0.004
Error	4	1370	343		
Total	5	14250			

Fit	StDev Fit	95.0% CI		95.0% PI	
-8.54	13.40	(-45.74,	28.67)	(-72.00,	54.92)
18.59	10.06	(-9.34,	46.52)	(-39.91,	77.10)
45.72	7.87	(23.85,	67.59)	(-10.15,	101.58)
72.85	7.87	(50.98,	94.72)	(16.98,	128.71)
99.98	10.06	(72.04,	127.91)	(41.47,	158.48)
127.10	13.40	(89.90,	164.31)	(63.65,	190.56)

Figure 6.5

The Minitab output, as shown in Figure 6.5, indicates that the predicted value of y for each given value of x (Fit), the standard error of prediction (Stdev. Fit), the 95% confidence interval (95% C.I.), and the 95% prediction interval (95% P.I.). For example, the third line down from the top corresponds to a temperature of 30. Thr predicted value is 45.72, the standard error of prediction is 7.87, the 95% confidence interval is 23.85 to 67.59, and the 95% prediction interval is -10.15 to 101.58. The data window stores the identical information from the session window in the columns following the data.

Minitab will also produce a fitted line plot with confidence and prediction intervals.

Follow this step to produce a fitted line plot with confidence and prediction intervals.
Choose **Stat**>**Regression**>**Fitted Line Plot.** Place Pressure in the
Response (**Y**): text box. Place Temp in the Predictor (**X**): text box.
Display the confidence and prediction bands.
Choose **Options.** Place checks in the **D**isplay confidence bands and Display prediction
bands Display options checkboxes.
Enter a title.
Type *Vapor Pressure of Water* in the **T**itle: text box. Choose **OK.**
Choose **OK.**

The Minitab Output

Vapor Pressure of Water

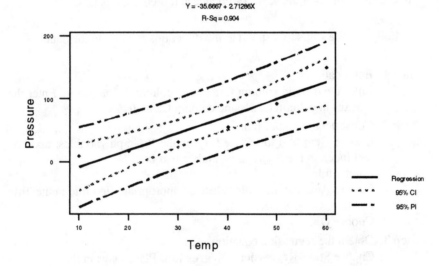

$Y = -35.6667 + 2.71286X$

R-Sq = 0.904

Figure 6.6

The Minitab output, as shown in Figure 6.6, contains a plot of the original data, the estimated regression line, the 95% confidence band, and the 95% prediction band. The confidence and the prediction bands both grow wider as the independent variable moves away from the center of the independent variable.

Exercises

6.1 (EXO6O1) Davidson (1993) studied the ozone levels in the South Coast Air Basin of California for the years 197 1991. He believes that the number of days the ozone levels exceeded 0.20 ppm (the response) depends on the seasonal meteorological index which is the seasonal average 850 millibar temperature (the independent variable). The data follow.

Year	Days	Index	Year	Days	Index
1976	91	16.7	1984	81	18.0
1977	105	17.1	1985	65	17.2
1978	106	18.2	1986	61	16.9
1979	108	18.1	1987	48	17.1
1980	88	17.2	1988	61	18.2
1981	91	18.2	1989	43	17.3
1982	58	16.0	1990	33	17.5
1983	82	17.2	1991	36	16.6

a. Make a scatterplot of the data.

b. Obtain the prediction equation.

c. Perform a complete, appropriate analysis. Discuss your results and conclusions.

d. Calculate and plot the 95% confidence and prediction bands.

Follow these steps to respond to the directions relating to the data.

Step 1. Enter data.
Enter the days in column C1. Name column C1 as *Days*. Enter the indexes in column C2. Name column C2 as *Index*.

Step 2. Construct the scatter plot.
Choose **Graph**>**Plot**. Place Days in the Y Graph variables: text box and Index in the X Graph variables: text box.
Enter a title.
Choose **A**nnotation>**T**itle. Place an appropriate first title in the Title text box. Choose **OK.**
Choose **OK.**

Step 3. Obtain the regression equation.
Choose **Stat**>**Regression**>**Regression.** Place Days in the Response: text box. Place Index in the Predictors: text box.
Display the confidence and prediction bands.
Choose Options. Place Index in the Prediction intervals for new observations: text box. Choose the default value of 95 (%) in the Confidence level: text box. Choose **OK.**
Choose **OK.**

Step 4. Construct the fitted line plot.
Choose **Stat**>**Regression**>**Fitted Line Plot.** Place Days in the Response (**Y**): text box. Place Index in the Predictor (**X**): text box.
Display the confidence and prediction bands.
Choose Options. Place checks in the **D**isplay **c**onfidence bands and Display prediction bands Display options checkboxes.
Enter a title.

Type *Ozone Levels* in the Title: text box. Choose **OK**.
Choose **OK**.

6.2 (EXO602) Mandel (1984) looks at the relationship between the amount of nickel (the independent variable) and the volume per cent austenite in various steels. The data follow.

x	y	x	y
0.608	2.11	0.710	2.51
0.634	1.95	0.730	2.33
0.651	2.27	0.750	2.26
0.658	1.95	0.772	2.47
0.675	2.05	0.802	2.80
0.677	2.09	0.819	2.95
0.702	2.54		

a. Make a scatterplot of the data.

b. Obtain the prediction equation.

c. Perform a complete, appropriate analysis. Discuss your results and conclusions.

d. Calculate and plot the 95% confidence and prediction bands.

6.3 (EXO603) Hsuie, Ma, and Tsai study the effect of the molar ratio of sebacic acid (the independent variable) on the intrinsic viscosity of copolyesters (the response). The data follow.

Ratio	Viscosity	Ratio	Viscosity
1.0	0.45	0.6	0.70
0.9	0.20	0.5	0.57
0.8	0.34	0.4	0.55
0.7	0.58	0.3	0.44

a. Make a scatterplot of the data.

b. Obtain the prediction equation.

c. Perform a complete, appropriate analysis. Discuss your results and conclusions.

d. Calculate and plot the 95% confidence and prediction bands.

6.4 (EXO604) Mehta and Deopura (1995) studied the mechanical properties of as spun PET-LCP blend fibers. They believe that the modulus (the response) depends on the % of PET in the blend. The data follow.

% PET	Modulus
100	2.12
97.5	2.26
95	2.57
90	3.26
80	3.46
50	4.54
0	8.5

a. Make a scatterplot of the data.

b. Obtain the prediction equation.

c. Perform a complete, appropriate analysis. Discuss your results and conclusions.

d. Calculate and plot the 95% confidence and prediction bands.

6.5 (EXO6O5) Byers and Williams (1987) studied the impact of temperature (the independent variable) on the viscosity (the response) of toluene-tetralin blends. The following data are for blends with a 0.4 molar fraction of toluene.

Temp. deg. C	Vis. $mPa \cdot s$
24.9	1.133
35.0	0.9772
44.9	0.8532
55.1	0.7550
65.2	0.6723
75.2	0.6021
85.2	0.5420
95.2	0.5074

a. Make a scatterplot of the data.

b. Obtain the prediction equation.

c. Perform a complete, appropriate analysis. Discuss your results and conclusions.

d. Calculate and plot the 95% confidence and prediction bands.

Follow these steps to perform the above operations.

Step 1. Enter data.

Enter the temperatures in column C1. Name column C1 as *Temp*. Enter the viscosities in column C2. Name column C2 as *Vis*.

Step 2. Construct the scatter plot.

Choose **Graph**>**Plot**. Place Vis in the Y Graph variables: text box and Temp in the X Graph variables: text box.

Enter a title.

Choose Annotation>Title. Place an appropriate first title in the Title text box. Choose **OK**.

Choose **OK**.

Step 3. Obtain the regression equation.

Choose **Stat**>**Regression**>**Regression**. Place Vis in the Response: text box. Place Temp in the Predictors: text box. Display the confidence and prediction bands.

Choose Options. Place Temp in the Prediction intervals for new observations: text box. Choose the default value of 95 (%) in the Confidence level: text box. Choose **OK.**
Choose **OK.**

Step 4. Construct the fitted line plot.
Choose **Stat>Regression>Fitted Line Plot.** Place Vis in the
Response (Y): text box. Place Temp in the Predictor (X): text box.
Display the confidence and prediction bands.
Choose Options. Place checks in the Display confidence bands and
Display prediction bands Display options checkboxes.
Enter a title.
Type *Impact of Temperature on Viscosity of Toluene-tetralin Blends* in
the Title: text box. Choose **OK.**
Choose **OK.**

6.6 (EXO608) Carroll and Spiegelman (1986) look at the relationship between the
pressure in a tank (the response) and the volume of liquid (the independent
variable). The data follow.

Vol.	Press.	Vol.	Press.	Vol.	Press.
2084	4599	2842	6380	3789	8599
2084	4600	3030	6818	3789	8600
2273	5044	3031	6817	3979	9048
2273	5043	3031	6818	3979	9048
2273	5044	3221	7266	4167	9484
2463	5488	3221	7268	4168	9487
2463	5487	3409	7709	4168	9987
2651	5931	3410	7710	4358	9936
2652	5932	3600	8156	4358	9938
2652	5932	3600	8158	4546	10377
2842	6380	3788	8597	4547	10379

a. Make a scatterplot of the data.

b. Obtain the prediction equation.

c. Perform a complete, appropriate analysis. Discuss your results and conclusions.

d. Calculate and plot the 95% confidence and prediction bands.

Follow these steps to perform the above operations.

Step 1. Enter data.
Enter the volumes in column C1. Name column C1 as *Volume*. Enter
the pressures in column C2. Name column C2 as *Pressure*.

Step 2. Construct the scatter plot.
Choose **Graph>Plot.** Place Pressure in the Y Graph variables: text
box and Volume in the X Graph variables: text box.
Enter a title.

Choose <u>A</u>nnotation><u>T</u>itle. Place an appropriate first title in the Title text box. Choose <u>O</u>K.

Choose <u>O</u>K.

Step 3. Obtain the regression equation.

Choose <u>Stat</u>><u>Regression</u>><u>Regression</u>. Place Pressure in the Re<u>s</u>ponse: text box. Place Volume in the Pre<u>d</u>ictors: text box. Display the confidence and prediction bands.

Choose <u>O</u>ptions. Place Volume in the Prediction <u>i</u>ntervals for new observations: text box. Choose the default value of 95 (%) in the Confidence <u>l</u>evel: text box. Choose <u>O</u>K.

Choose <u>O</u>K.

Step 4. Construct the fitted line plot.

Choose <u>Stat</u>><u>Regression</u>><u>F</u>itted Line Plot. Place Pressure in the Response (<u>Y</u>): text box. Place Volume in the Predictor (<u>X</u>): text box. Display the confidence and prediction bands.

Choose <u>O</u>ptions. Place checks in the <u>D</u>isplay <u>c</u>onfidence bands and Display <u>p</u>rediction bands Display options checkboxes.

Enter a title.

Type *Volume and Pressure* in the <u>T</u>itle: text box. Choose <u>O</u>K.

Choose <u>O</u>K.

6.4   Multiple Linear Regression

In most engineering problems, the response variable is a function of several in-dependent variables. Multiple linear regression is a straight foward extension of linear regression to more than one independent variable.

Multiple Linear Regression Model
Example 6.2 - Springs with Cracks
Engineering phenomena may be well modeled by a linear combination of several independent variables. Box and Bisgaard (1987) model the cracking of carbon-steel springs (the response) in terms of

1. the temperature of the steel before quenching,

2. the amount of carbon in the formulation, and

3. the temperature of the quenching oil.

The response variable in this study is the percentage of springs which do not exhibit cracking (y_i) for the ith batch produced. Let x_{i1} be the temperature of the steel for the ith batch produced. Let x_{i2} be the amount of carbon in the ith batch produced, and let x_{i3} be the temperature of the quenching oil for the ith batch produced.. The prediction equation might be written as

$$y_i = \beta_0 + \beta_1 x_{i1} + \beta_2 x_{i2} + \beta_3 x_{i3} + \epsilon_u$$

where

1. β_0 is the y-intercept,
2. β_1 is the coefficient associated with the steel temperature,
3. β_2 is the coefficient associated with the amount of carbon,
4. β_3 is the coefficient associated with the oil quenching temperature, and
5. ϵ_i is a random error.

The experimental results follow.

Run	Steel Temp.	Carbon	Oil Temp.	% Cracking
1	$1450^0 F$	0.50%	$70^0 F$	67
2	$1600^0 F$	0.50%	$70^0 F$	79
3	$1450^0 F$	0.70%	$70^0 F$	61
4	$1600^0 F$	0.70%	$70^0 F$	75
5	$1450^0 F$	0.50%	$120^0 F$	59
6	$1600^0 F$	0.50%	$120^0 F$	90
7	$1450^0 F$	0.70%	$120^0 F$	52
8	$1600^0 F$	0.70%	$120^0 F$	87

Follow these steps to determine the least squares multiple regression equation between the variables.

Step 1. Enter data.

Enter the steel temperatures in column C1. Name column C1 as *Steel*. Enter the amount of carbon in column C2. Name column C2 as *Carbon*. Enter the temperature of the quenching oil in column C3. Name column C3 as *Oil*. Enter the percentage of springs which do not exhibit cracking in column C4. Name column C4 as *Percent*.

Step 2. Obtain the regression equation.

Choose **Stat**>**Regression**>**Regression.** Place Percent in the Response: text box. Place Steel, Carbon and Oil in the Predictors: text box. Choose **OK**.

The Minitab Output

The regression equation is

Percent = - 150 + 0.153 Steel - 25.0 Carbon + 0.030 Oil

Predictor	Coef	StDev	T	P
Constant	-150.43	54.57	-2.76	0.051
Steel	0.15333	0.03375	4.54	0.010
Carbon	-25.00	25.31	-0.99	0.379
Oil	0.0300	0.1012	0.30	0.782

S = 7.159   R-Sq = 84.4%   R-Sq(adj) = 72.8%

Analysis of Variance

Source	DF	SS	MS	F	P
Regression	3	1112.50	370.83	7.24	0.043
Error	4	205.00	51.25		
Total	7	1317.50			

Source	DF	Seq SS
Steel	1	1058.00
Carbon	1	50.00
Oil	1	4.50

Figure 6.7

The Minitab output, as shown in Figure 6.7, indicates the regression equation between the percentage of springs which do not exhibit cracking and the predictor variables (the temperature of the steel before quenching, the amount of carbon in the formulation, and the temperature of the quenching oil). Since the p-value is less than $\alpha = .05$, we may reject the null hypothesis that there is no relationship between the response and independent variables ($H_0 : \beta_1 = \beta_2 = \beta_3 = 0$) and conclude that the percentage of springs which do not exhibit cracking does depend upon either the temperature of the steel before quenching, the amount of carbon in the formulation, or the temperature of the quenching oil. Further, we observe that $R^2 = 84.4\%$, which supports the idea that this is a reasonable model.

The Minitab output also indicates the least squares estimate of the intercept (Constant), the coefficient associated with the steel temperature (Steel), the coefficient associated with the amount of carbon (Carbon), and the coefficient associated with the oil quenching temperature (Oil). Beside each estimate is the appropriate standard deviation of the estimator. The calculated value of the test statistic and the p-value for each independent variable follow. We see that only the coefficient associated with the steel temperature (Steel) has a p-value less than $\alpha = .05$. As a result, we have evidence to suggest that only the steel temperature influences the percentage of springs which do not exhibit cracking, given the other independent variables in the model. Further, since the coefficient (0.15333) is positive, we see a higher fraction of springs without cracking as we increase the steel temperature.

Mintab will also produce confidence and prediction bands for the data.

Follow this step to determine confidence intervals and prediction intervals.
Choose **Stat**>**Regression**>**Regression.** Place Percent in the
Response: text box. Place Steel, Carbon and Oil in the Predictors: text box.
Obtain the prediction and confidence intervals.
Choose Options. Place Steel, Carbon and Oil in the Prediction intervals for new
observations: text box. Choose the default value of 95 (%) in the Confidence level:
text box. Choose **OK.**
Choose **OK.**

The Minitab Output

Fit	StDev Fit	95.0% CI		95.0% PI	
61.50	5.06	(47.44,	75.56)	(37.15,	85.85)
84.50	5.06	(70.44,	98.56)	(60.15,	108.85)
56.50	5.06	(42.44,	70.56)	(32.15,	80.85)
79.50	5.06	(65.44,	93.56)	(55.15,	103.85)
63.00	5.06	(48.94,	77.06)	(38.65,	87.35)
86.00	5.06	(71.94,	100.06)	(61.65,	110.35)
58.00	5.06	(43.94,	72.06)	(33.65,	82.35)
81.00	5.06	(66.94,	95.06)	(56.65,	105.35)

Figure 6.8

The Minitab output, as shown in Figure 6.8, indicates the 95% confidence and prediction bounds for the data used to generate the model for the spring data.

Exercises

6.7 (EXO611) Lawson (1982) conducted a designed experiment to optimize the
yield (the response) for a caros acid process. For proprietary reasons, he could
not discuss the three specific independent variables, x_1, x_2, and x_3. The data
follow in their actual run order. Perform a complete analysis.

Run	x_1	x_2	x_3	y	Run	x_1	x_2	x_3	y
1	-1	1	-1	77	15	1	1	-1	80
2	1	1	1	92	16	-1	-1	-1	68
3	0	0	0	81	17	-1	1	1	92
4	-1	-1	1	86	18	0	0	0	81
5	1	-1	-1	67	19	0	0	0	80
6	0	0	0	82	20	1	-1	1	86
7	0	0	-1	72	21	0	-1	0	76
8	-1	0	0	84	22	0	0	-1	71
9	0	1	0	81	23	-1	0	0	86
10	0	0	1	87	24	0	0	0	82
11	0	0	0	82	25	0	0	0	81
12	0	-1	0	74	26	0	0	1	86
13	0	0	0	82	27	0	1	0	82
14	1	0	0	78	28	1	0	0	79

Follow these steps to determine the least squares multiple regression equation between the variables.

Step 1. Enter data.

Enter the runs in column C1. Name column C1 as *Runs*. Enter the independent variables in columns C2, C3 and C4. Name column C2 as *x1*, column C3 as *x2* and column c4 as *x3*. Enter the yields (the response) in column C5. Name column C5 as *y*.

Step 2. Obtain the regression equation.

Choose **Stat**>**Regression**>**Regression.** Place y in the Response: text box. Place x1, x2 and x3 in the Predictors: text box. Obtain the prediction and confidence intervals.

Choose **O**ptions. Place x1, x2 and x3 in the Prediction intervals for new observations: text box. Choose the default value of 95 (%) in the Confidence level: text box. Choose **OK.**

Choose **OK.**

6.8 (EXO610) Tracy, Young, and Mason studied the impact of temperature, x_1, and concentration, x_2, on the percentage of impurities, y, for a chemical process. The data follow. Perform a complete analysis.

Temp.	Conc.	% Impurities	Temp.	Conc.	% Impurities
85.8	42.3	14.9	85.9	43.4	16.9
83.8	43.4	16.9	85.7	43.3	16.7
84.5	42.7	17.4	86.3	42.6	16.9
86.3	43.6	16.9	83.5	44.0	16.7
85.2	43.2	16.9	85.8	42.8	17.1
83.8	43.7	16.7	85.9	43.1	17.6
86.1	43.3	17.1	84.2	43.5	16.9

6.9 (EXO613) Wauchope and McDowell (1984) studied the effect of the amount of extractable iron, the amount of extractable aluminum, and the pH of soils on the soils' adsorption of phosphate. The data follow. Perform a complete analysis.

Extractable Iron	Extractable Aluminum	pH	Adsorption Index
61	13	7.7	4
175	21	7.7	18
111	24	6.8	14
124	23	7.3	18
130	64	5.1	26
173	38	5.7	26
169	33	5.8	21
169	61	5.2	30
160	39	6.3	28
244	71	5.7	36
257	112	4.4	65
333	88	4.5	62
199	54	6.2	40

Follow these steps to determine the least squares multiple regression equation between the variables.

Step 1. Enter data.

Enter the extractable iron in column C1. Name column C1 as *Iron*. Enter the extractable aluminum in column C2. Name column C2 as *Aluminum*. Enter the pH in column C3. Name column C3 as *pH*. Enter the adsorption indexes in column C4. Name column C4 as *Index*.

Step 2. Obtain the regression equation.

Choose **Stat>Regression>Regression.** Place Index in the Response: text box. Place Iron, Aluminum and pH in the Predictors: text box.

Obtain the prediction and confidence intervals.

Choose **Options.** Place Iron, Aluminum and pH in the Prediction intervals for new observations: text box. Choose the default value of 95 (%) in the Confidence level: text box. Choose **OK.** Choose **OK.**

6.10 (EXO612) Said et al. (1994) studied the effect the mole contents of cobalt, x_1, and calcination temperature, x_2, on the surface area of an iron-cobalt hydoxide catalyst. The data follow. Perform a complete analysis.

x_1	x_2	y	x_1	x_2	y
0.6	200	90.6	2.6	200	53.1
0.6	250	82.7	2.6	250	52.0
0.6	400	58.7	2.6	400	43.4
0.6	500	43.2	2.6	500	42.4
0.6	600	25.0	2.6	600	31.6
1.0	200	127.1	2.8	200	40.9
1.0	250	112.3	2.8	250	37.9
1.0	400	19.6	2.8	400	27.5
1.0	500	17.8	2.8	500	27.3
1.0	600	9.1	2.8	600	19.0

6.5   Residual Analysis

The ability of the model that we are build to predict data determines the usefullness of the model. The residuals, the differences $(y - \hat{y})$ between the observed values and their corresponding predicted values, provide a measure of how well the model predicts the data. These residuals represent the failure of the model to predict the given data. An analysis of these residuals provides insight as to how well our model estimates the data.

Two plots are useful when we check for outliers or an inadequate model:

(i) a plot of the residuals against the predicted values, and

(ii) a plot of the residuals against the independent variables.

Outliers appear as distinctly different values in the plot of the residuals against the predicted values.

A clear pattern in the plot of the residuals against the independent variables is indicative of an inadequate model. If an adequate model is fitted to data, there will be no clear pattern to the residuals.

Residual Analysis in Identifying Outliers
Example 6.3 - Relationship of Hardness and
Young's Modulus for High Density Penetrator Materials
The military often uses high density materials in metals in munitions since these metals have a better ability to penetrate armor. Magness (1944) used an observational study of seven different high density metals by measuring the Rockwell hardness (the independent variable) and Young's modulus (the response).

Hardness	Young's Modulus
41	310
41	340
44	380
40	317
43	413
15	62
40	119

Follow these steps to determine the least squares equation between Young's modulus and hardness.

Step 1. Enter data.

Enter the hardness in column C1. Name column C1 as *Hardness*. Enter Young's modulus in column C2. Name column C2 as *Modulus*.

Step 2. Obtain the regression equation.

Choose **Stat**>**Regression**>**Regression.** Place Modulus in the Response: text box. Place Hardness in the Predictors: text box. Store the residuals and predicted values (FITS).

Choose Storage. Place a check in the Residuals Storage checkbox. Place a check in the Fits Storage checkbox. Choose **OK.** (The residuals will be stored in a column named RESI1, and the fits will be stored in a column named FITS1.)

Choose **OK.**

The Minitab Output

Regression Analysis

```
The regression equation is
Modulus = - 109 + 10.2 Hardness

Predictor        Coef       StDev         T         P
Constant        -108.5       144.5      -0.75     0.487
Hardness        10.230       3.719       2.75     0.040

S = 92.25       R-Sq = 60.2%      R-Sq(adj) = 52.3%

Analysis of Variance

Source        DF          SS          MS         F         P
Regression     1        64401       64401      7.57     0.040
Error          5        42550        8510
Total          6       106951
```

Figure 6.9

A Minitab analysis of the data, as shown in Figure 6.9, indicates that the Rockwell hardness does predict the Young's modulus. The R^2 value indicates either a physical system with a lot of inherent variability or some room for improvement with our model.

Follow this series of steps to construct:

(A) a boxplot of the residuals

(B) a plot of the residuals against the predicted values

(C) a plot of the residuals against the independent variable.

Step 1. Construct the boxplot of the residuals stored in RESI1.

Choose **Graph**> **Boxplot**. Place RESI1 in the Graph Variables: text box. Choose **OK**.

The Minitab Output

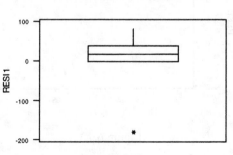

Figure 6.10

The boxplot produced by Minitab, as shown in Figure 6.10, indicates an outlier, which is the seventh observation.

151

Step 2. Obtain the plot of the residuals against the predicted values.
   Choose **Stat**>**Regression**>**Residual Plots.** Place RESI1 in the
   Residuals: text box. Place FITS1 in the Fits: text box.
   Enter a title.
   Place Young's Modulus vs. Hardness in the Title: textbox. Choose **OK.**
Step 3. Obtain the plot of the residuals against the independent variable.
   Choose **Graph**>**Plot**. Place RES1 in the Y Graph variables: text box and
   Hardness in the X Graph variables: text box.
   Enter a title.
   Choose Annotation>Title. Place an appropriate first title in the Title text
   box. Choose **OK.**
   Choose **OK.**

The Minitab Output

Observation Number

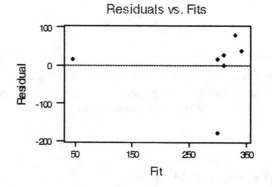

Figure 6.11

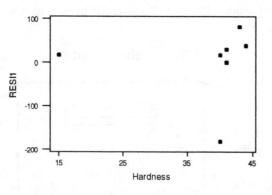

Figure 6.12

Looking at the Mintab graphics output, as shown in Figure 6.11, we see that the
plot of the residuals against the predicted values is the plot in the lower left hand
corner. Again, this plot indicates that the model does not predict the seventh ob-

servation. Looking at the Mintab graphics output again, we see that the plot of the residuals against the independent variable is the plot in the lower left hand corner. Again, this plot indicates that the model does not predict the seventh observation. The Mintab output, as shown in Figure 6.12, again indicates that the model does not predict the seventh observation.

Model Misspecification and Residual Analysis
Example 6.4 - Vapor Pressure of Water - Continued

Chemical and mechanical engineers often need to know the vapor pressure of water for specific temperatures. One possible approach is to use a simple model to predict the vapor pressure given the temperature. You looked at this model in Example 6.1, and we will continue the analysis of the model in this example. Figure 6.1 contains the scatterplot of the data and Figure 6.2 indicates the Mintab analysis of the Vapor Pressure Data. Observe, in Figure 6.2, that R^2 is 90.4% and R^2_{adj} is 88.0%. The simple regression model may be adequate for predicting the vapor pressures by temperatures; however, since we are dealing with physical phenomena, we may be able to generate a better model.

Temp. Deg. C	Vapor Pres. mm Hg
10	9.2
20	17.5
30	31.8
40	55.3
50	92.5
60	149.4

Follow this series of four steps to perform a residual analysis of the data.

Step 1. Enter data.

Enter the temperatures in column C1. Name column C1 as *Temp*. Enter the vapor pressures in column C2. Name column C2 as *Pressure*.

Step 2. Obtain the regression equation.

Choose **Stat**>**Regression**>**Regression.** Place Pressure in the Response: text box. Place Temp in the Predictors: text box. Store the residuals and predicted values (FITS).

Choose Storage. Place a check in the Residuals Storage checkbox. Place a check in the Fits Storage checkbox. Choose **OK.** (The residuals will be stored in a column named RES1, and the fits will be stored in a column named FITS1.)

Choose **OK.**

Step 3. Obtain the plot of the residuals against the predicted values.

Choose **Graph**>**Plot**. Place RES1 in the Y Graph variables: text box and FITS1 in the X Graph variables: text box.

Enter a title.

Choose Annotation>Title. Place an appropriate first title in the Title text

box. Choose **OK.**
Choose **OK.**

The Minitab Output

Residuals Versus Predicted Values
for the Vapor Pressure Data

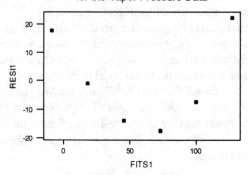

Figure 6.13

The Minitab output, as shown in Figure 6.13, indicates that the residuals for low temperatures are positive, for moderate temperatures they are negative, and for high temperatures, they are once again positive. The pattern appears quadratic.

Step 4. Obtain the plot of the residuals against temperature.

Choose **Graph**>**Plot**. Place RES1 in the Y **G**raph variables: text box and Temp in the X **G**raph variables: text box.
Enter a title.
Choose **A**nnotation>**T**itle. Place an appropriate first title in the Title text box. Choose **OK.**
Choose **OK.**

The Minitab Output

Residuals Versus Temperature
for the Vapor Pressure Data

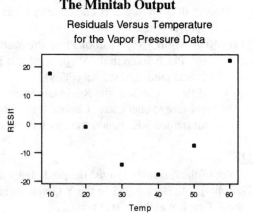

Figure 6.14

The Minitab output, as shown in Figure 6.14, reveals the same pattern, suggesting that a better model should include a quadratic term for temperature or some other method to account for the curvature in the data.

Checking Assumptions

Least squares estimation of the model assumes that the residuals
> **1** have an expected value of zero
> **2** have constant variance, and
> **3** are independent.

Testing procedures require the additional assumption that the residuals require a well behaved distribution.
Graphical techniques useful for checking assumptions include:

- a plot of the residuals against the predicted values, which checks the constant variance assumption,
- a plot of the residuals against the independent variable, which also checks the constant variance assumption,
- a plot of the residuals in time order, which checks the independce assumption,
- a stem-and-leaf plot of the residuals, which checks the well-behaved distributed assumption, and
- a normal probability plot, which also checks the well behaved distribution assumption.

Statisticians tend to focus the most attention on the constant variance assumption. In many engineering problems, the variability increases with the predicted value and the resulting residual plot displays a distinct funnel effect. A transformation of either the response or the independent variables may be appropriate to correct this problem.
A plot of the residuals in time order is appropriate to ensure that no systematic biases occur over the course of the data collection.
The stem-and-leaf plot and the normal probability plot provide methods for checking the shape of the distribution of the residuals. Minitab provides a normal probability plot. If the data truly come from a normal distribution, the resulting plot should look like a straight line. Deviations indicate an evidence of non-normality: the bigger the deviation, the bigger the problem.

Residual Analysis with a Multiple Regression Model
Example 6.5 - Popcorn Data
A popular student project looks at making popcorn. The following data represent an attempt to find the optimal combination of burner setting (x_1), amount of oil

(x_2), and "popping" time (x_3) to minimize the number of inedible kernels (y) when popping the corn over a stove top. A possible model might be

$$y_i = \beta_0 + \beta_1 x_{i1} + \beta_2 x_{i2} + \beta_3 x_{i3} +$$
$$\beta_{11} x_{i1}^2 + \beta_{22} x_{i2}^2 + \beta_{33} x_{i3}^2 +$$
$$\beta_{12} x_{i1} x_{i2} + \beta_{13} x_{i1} x_{i3} + \beta_{23} x_{i2} x_{i3} + \epsilon$$

The following data summarize the experiment.

Temp	Oil	Time	y	Temp	Oil	Time	y
7	4	90	24	5	4	90	32
5	3	105	28	6	2	75	32
7	3	105	40	5	2	90	34
7	2	90	22	7	3	75	17
6	4	105	11	6	3	90	30
6	3	90	16	6	3	90	17
5	3	75	126	6	4	75	50
6	2	105	34				

Follow these steps to determine the least squares multiple regression equation between the variables.

Step 1. Enter data.

Enter the temperatures in column C1. Name column C1 as *X1*. Enter the amount of oil in column C2. Name column C2 as *X2*. Enter the "popping" time in column C3. Name column C3 as *X3*. Enter the number of inedible kernels (y) in column C4. Name column C4 as *Y*.

Step 2. Introduce additional variables.

Introduce $X1^2$.

Choose **Calc**>**Calculator**. Type *X11* in the Store result in variable: text box. Place X1**2 in the Expression: text box. Choose **OK.**

Introduce $X2^2$.

Choose **Calc**>**Calculator**. Type *X22* in the Store result in variable: text box. Place X2**2 in the Expression: text box. Choose **OK.**

Introduce $X3^2$.

Choose **Calc**>**Calculator**. Type *X33* in the Store result in variable: text box. Place X3**2 in the Expression: text box. Choose **OK.**

Introduce $X1 \cdot X2$.

Choose **Calc**>**Calculator**. Type *X12* in the Store result in variable: text box. Place X1*X2 in the Expression: text box. Choose **OK.**

Introduce $X1 \cdot X3$.

Choose **Calc**>**Calculator**. Type *X13* in the Store result in variable: text box. Place X1*X3 in the Expression: text box. Choose **OK.**

Introduce $X2 \cdot X3$.

Choose **Calc**>**Calculator**. Type *X23* in the Store result in variable: text box. Place X2*X3 in the Expression: text box. Choose **OK.**

Step 3. Obtain the regression equation.

Choose **Stat**>**Regression**>**Regression.** Place Y in the
Response: text box. Place X1, X2, X3, X11, X22, X33, X12, X13, and
X23 in the Predictors: text box.

Store the predicted values and the residuals.

Choose Storage Place checks in the Fits and Residuals checkboxes. Place
a check in the Fits Storage checkbox. Choose **OK.**

Choose **OK.**

The Minitab Output

Regression Analysis

```
The regression equation is
Y = 2122 - 380 X1 + 109 X2 - 23.2 X3 + 16.5 X11 - 4.50 X22 + 0.0678 X:
          - 4.00 X12 + 2.02 X13 - 0.683 X23
```

Predictor	Coef	StDev	T	P
Constant	2122.1	619.7	3.42	0.019
X1	-379.6	123.2	-3.08	0.027
X2	109.37	91.65	1.19	0.286
X3	-23.183	8.214	-2.82	0.037
X11	16.500	9.034	1.83	0.127
X22	-4.500	9.034	-0.50	0.640
X33	0.06778	0.04015	1.69	0.152
X12	-4.000	8.680	-0.46	0.664
X13	2.0167	0.5786	3.49	0.018
X23	-0.6833	0.5786	-1.18	0.291

```
S = 17.36      R-Sq = 85.5%      R-Sq(adj) = 59.3%
```

Analysis of Variance

Source	DF	SS	MS	F	P
Regression	9	8869.0	985.4	3.27	0.103
Error	5	1506.7	301.3		
Total	14	10375.7			

Figure 6.15

The analysis of variance, as shown in Figure 6.15, indicates a marginal model since
the p-value is 0.103. The R^2 value of 85.5% indicates that the model accounts
over 85% of the observed variablity. The R^2_{adj} value of 59.3% indicates that the
model is possibly over specified. Only the coefficients associated with X1 (Temp),
X3 (Time) and X13 (the interaction of time and temperature) have p-values less
than 0.05 and therefore appear significant. none of the terms involving X2 (Oil),
appears to be important.

Follow these two steps to perform an analysis of the residuals.

Step 1. Obtain the residual plots.

Choose **Stat**>**Regression**>**Residual Plots.** Place RESI1 in the Residuals: text box. Place FITS1 in the Fits: text box.

Enter a title.

Place Popcorn Data in the Title: textbox. Choose **OK.**

The Minitab Output

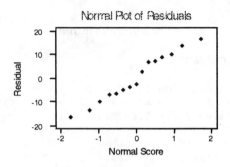

Figure 6.16

The Minitab Output

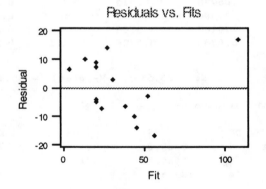

Figure 6.17

The normal probability plot, as shown in Figure 6.16, appears to be S-like. This pattern suggests some kind of transformation is necessary. The plot of the residuals against the fits, as shown in Figure 6.17, suggests that variability may increase with the predicted value. This pattern may justify using a square root or log transfomation of the response and reanalyzing the data.

Step 2. Obtain the plot of the residuals against the independent variable.

Choose **Graph**>**Plot**. Place RES1 in the Y Graph variables: text box and Time in the X Graph variables: text box.

Enter a title.

Choose <u>A</u>nnotation><u>T</u>itle. Place an appropriate first title in the Title text box. Choose **OK.**
Choose **OK.**

The Minitab Output

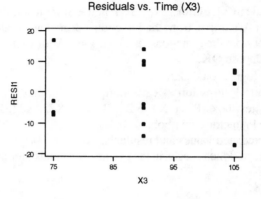

Figure 6.18

The plot of the residuals against the independent variable time (X3), as shown in Figure 6.18, supports the assumption that there are no systematic effects over the course of the experiment and that the independence assumption is valid. Additional plots continue to support the validity of the assumptions.

Transformations

Engineers use transformations of the response for two fundamental reasons:
> **1** to correct problems with underlying assumptions, and
> **2** to change the natural metric of the problem in accord with engineering theory.

Typically, a transformation of the response is appropriate when problems are encountered with the constant variance assumption or when the residuals appear to follow a distinctly nonnormal distribution. The square root transformation is often used with count data, such as the popcorn data.

A Square Root Transformation
Example 6.6 - Popcorn Data - Continued

Follow these steps to perform a square root transformation and reanalyze the data.

Step 1. Introduce the square root transformation.

        Choose **Calc**>**Calculator**. Type *Sqrt* in the **S**tore result in variable: text box. Select Arithmetic from the **F**unctions: drop down dialog box. Select Square root from the **F**unctions: list box. Place Sqrt(Y) in the **E**xpression: text box. Choose **OK.**

Step 2. Obtain the regression equation.

        Choose **Stat**>**Regression**>**Regression.** Place Sqrt in the **R**esponse: text box. Place X1, X2, X3, X11, X22, X33, X12, X13, and X23 in the Pre**d**ictors: text box.

        Store the predicted values and residuals.

        Select **S**torage. Place checks in the **F**its and **R**esiduals Storage checkboxes. Choose **OK.**

        Choose **OK.**

The Minitab Output

Regression Analysis

```
The regression equation is
Sqrt = 144 - 26.7 X1 + 7.91 X2 - 1.46 X3 + 1.23 X11 - 0.044 X22 +
       - 0.352 X12 + 0.136 X13 - 0.0655 X23
```

Predictor	Coef	StDev	T	P
Constant	144.00	38.59	3.73	0.014
X1	-26.652	7.672	-3.47	0.018
X2	7.913	5.707	1.39	0.224
X3	-1.4618	0.5114	-2.86	0.035
X11	1.2278	0.5625	2.18	0.081
X22	-0.0444	0.5625	-0.08	0.940
X33	0.004355	0.002500	1.74	0.142
X12	-0.3519	0.5404	-0.65	0.544
X13	0.13558	0.03603	3.76	0.013
X23	-0.06548	0.03603	-1.82	0.129

```
S = 1.081      R-Sq = 87.8%      R-Sq(adj) = 65.8%
```

Analysis of Variance

Source	DF	SS	MS	F	P
Regression	9	42.001	4.667	3.99	0.071
Error	5	5.842	1.168		
Total	14	47.842			

Figure 6.19

The overall analysis of variance, as shown in Figure 6.19, provides a little more evidence that at least one of the coefficients of the variables is nonzero, although it is still not significant relative to the .05 significance level. The R^2 value of 87.8% indicates that the model accounts over 87% of the observed variablity. None of the terms associated with X2 (Time) are significant and that suggests that we drop all such terms from the model.

Follow this series of two steps to perform a reanalysis with X2, X22, X12 and X23 deleted from the model.

Step 1. Obtain the regression equation.

Choose **Stat>Regression>Regression.** Place Sqrt in the
Response: text box. Place X1, X3, X11, X33, and X13 in the Predictors: text box.
Store the predicted values and residuals.
Select Storage. Place checks in the Fits and Residuals Storage checkboxes.
Choose **OK.**
Choose **OK.**

The Minitab Output

Regression Analysis

```
The regression equation is
Sqrt = 168 - 27.7 X1 - 1.66 X3 + 1.23 X11 + 0.00437 X33 + 0.136 X13

Predictor        Coef        StDev           T           P
Constant       167.56        36.33        4.61       0.000
X1            -27.749         7.728       -3.59       0.006
X3            -1.6610         0.5152      -3.22       0.010
X11            1.2312         0.5794       2.12       0.063
X33          0.004370       0.002575       1.70       0.124
X13           0.13558        0.03722       3.64       0.005

S = 1.117      R-Sq = 76.5%      R-Sq(adj) = 63.5%

Analysis of Variance

Source        DF          SS          MS          F          P
Regression     5      36.620       7.324       5.87      0.011
Error          9      11.222       1.247
Total         14      47.842
```

Figure 6.20

This analysis of variance, as shown in Figure 6.20, does indicate that the overall regression is significant and that at least one of the independent variables does influence the number of inedible kernels. The temperature (X1), the popping time (X3), and their interaction (X13) are all significant. The two quadratic terms appear to be less important. We need to check the residuals.

Follow this second step in the series to perform an analysis of the residuals.

Step 2. Obtain the residual plots.

Choose **Stat**>**Regression**>**Residual Plots.** Place RESI3 in the
Residuals: text box. Place FITS3 in the Fits: text box.
Enter a title.
Place Popcorn Data in the Title: textbox. Choose **OK.**

The Minitab Output

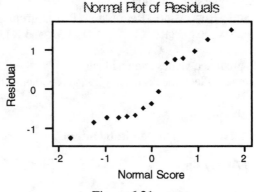

Figure 6.21

The Minitab Output
Observation Number

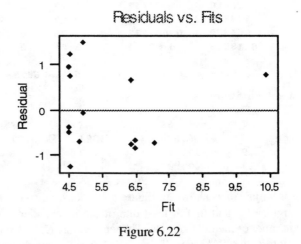

Figure 6.22

The normal probability plot, as shown in Figure 6.21, indicates that the transformation has increased the S-like pattern. This is indicates that the transformation
did not improve the distributional assumptions. The residual analysis involving a
plot of the residual against the predicted values, as shown in Figure 6.22, indicates

that we no longer see that variability increased with predicted value. There appears to be no apparent pattern to the residuals, indicating that the transformation has eliminated this problem. On the whole, the square root transformation appears to provide a better basis for modeling these data. We can conclude that a reasonable model for the data is

$$\sqrt{y} = 168 - 27.7x_1 - 1.66x_3 + 1.23x_{11} + 0.00437x_{33} + 0.136x_{13} + \epsilon.$$

A Logarithmic Transformation
Example 6.6 - Vapor Pressure - Continued
Physical chemistry suggests that the vapor pressure should follow an exponential relationship in the inverse of the temperature. Specifically, let p_v be the vapor pressure, and let T be the temperature. The Clausius-Clapeyron equation states that

$$\ln(p_v)\alpha - \frac{1}{T}.$$

Let y_i be the natural log of the ith vapor pressure, and let x_i be the inverse of the ith temperature. The Clausius-Clapeyron equation suggests that a reasonable model for the vapor pressure over a wide range of temperatures is

$$y_i = \beta_0 + \beta_1 x_i + \epsilon_i.$$

The following table summarizes the vapor pressures of water from 0^0C to 100^0C. The temperatures have been changed from degrees C to degrees K by adding 273 to each temperature.

Temp. Deg. K	Vapor Pres. mm Hg	Temp. Deg. K	Vapor Pres. mm Hg
273	4.6	333	149.4
283	9.2	343	233.7
293	17.5	353	355.1
303	31.8	363	525.8
313	55.3	373	760.0
323	92.5		

Follow these steps to analyze the data.

Step 1. Enter data.

Enter the temperatures in column C1. Name column C1 as *Temp*. Enter the vapor pressure in column C2. Name column C2 as *Pressure*.

Step 2. Construct a scatter plot of the original data.

Choose **Graph>Plot**. Place Pressure in the Y **G**raph variables: text box and Temp in the X **G**raph variables: text box.

Enter a title.

Choose **A**nnotation>**T**itle. Place an appropriate first title in the Title text box. Choose **OK**.

Choose **OK**.

The Minitab Output

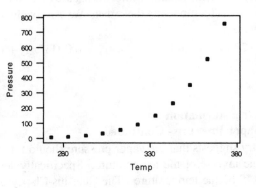

Vapor Data - Original Data

Figure 6.23

The scatter plot, as shown in Figure 6.23, indicates an exponential function. Theory from physical chemistry suggests that vapor pressure is an exponential function of the inverse of the temperature. We often use log transformation when we wish to linearize exponential data.

Follow these steps to transform the data.

Step 1. Introduce transformed variables.

Introduce the inverse of temperature.

Choose **Calc**>**Calculator**. Type *TempInv* in the Store result in variable: text box. Type *1/Temp* in the Expression: text box. Choose **OK.**

Introduce the natural logarithm of pressure.

Choose **Calc**>**Calculator**. Type *NatLog* in the Store result in variable: text box. Place Loge (Pressure) in the Expression: text box. Choose **OK.**

Step 2. Obtain the regression equation.

Choose **Stat**>**Regression**>**Regression.** Place NatLog in the Response: text box. Place TempInv in the Predictors: text box.
Store the predicted values and the residuals.
Choose Storage. Place checks in the Fits and Residuals Storage checkboxes. Choose **OK.**
Choose **OK.**

The Minitab Output

Regression Analysis

```
The regression equation is
NatLog = 20.6 - 5201 TempInv

Predictor        Coef        StDev           T         P
Constant      20.6074       0.0633      325.79     0.000
TempInv      -5200.76        20.14     -258.29     0.000

S = 0.02067     R-Sq = 100.0%     R-Sq(adj) = 100.0%

Analysis of Variance

Source       DF         SS          MS          F        P
Regression    1     28.511      28.511   66715.48    0.000
Error         9      0.004       0.000
Total        10     28.515
```

Figure 6.24

The estimated model, as shown in Figure 6.24, provides an excellent fit to the data as evidenced by an R^2 of 100.0% and p-values for all tests near 0.

Follow this step to obtain a plot of the transfomed data.
Choose **Graph**>**Plot**. Place NatLog in the Y Graph variables: text box and Temp-Inv in the X Graph variables: text box.
Enter a title.
Choose Annotation>Title. Place an appropriate first title in the Title text box.
Choose **OK.**
Choose **OK.**

The Minitab Output

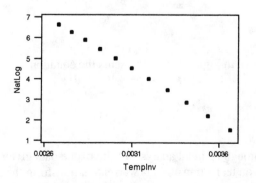

Figure 6.25

165

The fitted values (FITS1) are the predicted values using the equation

$$NatLog = 20.6 - 5201 TimeInv.$$

Follow these steps to convert the transformed data back to the original metric and plot the non-linear equation.

Step 1. Convert NatLog data.

Choose **Calc>Calculator**. Type *NewY* in the Store result in variable: text box. Choose Logarithm from the Functions: drop down dialog box. Select Exponentiate from the Functions: drop down list box. Place Expo (FITS1) in the Expression: text box. Choose **OK.**

Step 2. Obtain the plot of the non-linear equation.

Choose **Graph>Plot**. Place NewY in the Y Graph variables: text box and Temp in the X Graph variables: text box.

Connect the ordered pairs.

Click on the Data Display: Display drop down dialog box. Choose Connect.

Enter a title.

Choose **Annotation>Title**. Place an appropriate first and second title in the Title text box. Choose **OK.**

Choose **OK.**

The Minitab Output

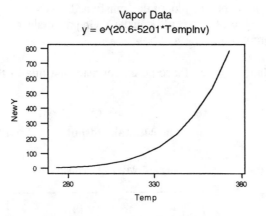

Figure 6.26

The plot, as shown in Figure 6.26, represents the equation

$$y = e^{20.6 - 5201 TempInv}$$

Exercises

6.11 Perform thorough residual analysis of the ozone data given in Exercise 6.1 of the text. If you feel a transition is warrented, transform the data and reanalyze the data. Discuss your results and conclusions. Please note that these data appear in time order.

6.12 Perform thorough residual analysis of the steel data given in Exercise 6.2 of the text. If you feel a transition is warrented, transform the data and reanalyze the data. Discuss your results and conclusions. Please note that these data appear in time order.

6.13 Perform thorough residual analysis of the springs with cracks data given in Example 6.11 of the text. If you feel a transition is warrented, transform the data and reanalyze the data. Discuss your results and conclusions.

6.14 Perform thorough residual analysis of the PET data given in Exercise 6.4 of the text.

6.15 Perform thorough residual analysis of the caros acid data given in Exercise 6.11 of the text. If you feel a transition is warrented, transform the data and reanalyze the data. Discuss your results and conclusions. Please note that these data appear in time order.

6.16 Perform thorough residual analysis of the catalyst data given in Exercise 6.12. If you feel a transition is warrented, transform the data and reanalyze the data. Discuss your results and conclusions.

6.17 Perform thorough residual analysis of the soil adsorption data given in Exercise 6.13 of the text. If you feel a transition is warrented, transform the data and reanalyze the data. Discuss your results and conclusions.

6.18 From basic principles of physical chemistry, viscosity is an exponential function of the temperature. Use appropriate transformations of the viscosity data given in Exercise 6.5 and perform a thorough residual analysis.

Chapter 7
Designing Experiments

Vining's text has developed basic concepts in collecting data, designing experiments, data analytic techniques, simple estimation and testing processes, and model building. We now undertake the construction and analysis of engineering experiments.

7.1   The 2^2 Factorial Design

New Minitab Commands (and some Minitab commands used previously)

1. **Stat>Regression>Regression** - Performs simple, polynomial regression, and multiple regression using the least squares method. In this section, you will use this command to obtain the regression equation for the data in Example 7.1 -Burst Strength of Packages.

2. **Stat>ANOVA>Interactions Plot** - Draws a single interaction plot if 2 factors are entered, or a matrix of interaction plots if more than 2 factors are entered. In this section, you will obtain an interactions plot to illustrate the presence or absence of an interaction.

Two-level factorial experiments are an initial approach to experimental design.

Example 7.1 - Burst Strength of Packages
Biodynamics, International ships biological samples across the nation. The FDA required an experiment to examine the sealing for the packages containing the samples. Of particular concern was the "burst" strength of a sealed package. Two temperatures controlling this strength were varied. For Temperature 1, the levels were $220^0 F$ and $230^0 F$. For Temperature 2, the levels were $120^0 F$ and $130^0 F$.

The Model and Calculating Effects
For this situation, the model is
$$y_i = \beta_0 + \beta_1 x_{i1} + \beta_2 x_{i2} + \beta_{12} x_{i1} x_{i2} + \epsilon_u$$
where

(A) y_i is the burst strength for the ith package tested,

(B) $x_{i1} = \begin{cases} \text{-1 if Temp. 1 is } 220^0 F \\ \text{1 if Temp. 1 is } 230^0 F \end{cases}$,

(C) $x_{i2} = \begin{cases} \text{-1 if Temp. 1 is } 120^0 F \\ \text{1 if Temp. 1 is } 130^0 F \end{cases}$,

(D) β_0 is the y-intercept,

(E) β_1 is the coefficient associated for Temperature 1,

(F) β_2 is the coefficient associated for Temperature 2,

(G) β_{12} is the interaction coefficient for Temperature 1 and Temperature 2, and

(H) ϵ_i is a random error.

We can estimate the effects of Temperature 1, Temperature 2, and their interaction. Let a denote the average of the responses when A (Temperature 1) is at its high level and B is at its low level. Similarly, let b denote the average of the responses when B (Temperature 2) is at its high level and A is at its low level. By following this convention, ab is the average of the responses when both A and B are at their high levels. By convention, we let (1) represent the average of the responses when all the factors are at their low levels. At this point, we are more interested in the effects and interactions conceptually than computationally. The experimental results follow.

x_1	x_2	y_i
-1	-1	2.76
1	-1	2.52
-1	1	2.74
1	1	2.70
-1	-1	2.78
1	-1	2.52
-1	1	2.70
1	1	2.56

Follow these four steps to estimate the effects and to determine the least squares multiple regression equation between the variables.

Step 1. Enter data.

Enter the temperature codes for Temperature 1 in column C1. Name column C1 as *X1*. Enter the temperature codes for Temperature 2 in column C2. Name column C2 as *X2*. Enter the "burst" strengths ($y_i's$) in column C3. Name column C3 as *Y*.

Step 2. Introduce the interaction variable.

Choose **Calc**>**Calculator**. Type *X1X2* in the <u>S</u>tore result in variable: text box. Choose Statistics from the <u>F</u>unctions: drop down dialog box. Select Sum from the <u>F</u>unctions: list box. Place Sum(X1*X2) in the <u>E</u>xpression: text box. Choose **OK**.

Step 3. Estimate the effects.

(See the text for an explanation for these equations.) Choose **Calc**>**Calculator**. Type *A* in the <u>S</u>tore result in variable: text box. Place Sum(X1*Y)/4 in the <u>E</u>xpression: text box. Choose **OK**.

Choose **Calc**>**Calculator**. Type *B* in the <u>S</u>tore result in variable: text box. Place Sum(X2*Y)/4 in the <u>E</u>xpression: text box. Choose **OK**.

Choose **Calc**>**Calculator**. Type *AB* in the <u>S</u>tore result in variable: text box. Place Sum(X1X2*Y)/4 in the <u>E</u>xpression: text box. Choose **OK**.

Use {Ctrl} D to go to the Data window. Under A we see that the main effect of Temp. 1 (A) is -0.17, and that the main effect of Temp. 2 (B) is -0.030. The main effect for Temperature 1 is much larger than the main effect for Temperature 2, which suggests that the burst strength depends more on Temperature 1 than on Temperature 2. In addition, we see that Temperature 1 has a negative effect on the burst strength. As a result, the higher we go on Temperature 1, the weaker the burst strength.

Under AB we see that the interaction effect (AB) is 0.08 and is consistent with the size of the main effects. We thus see a positive interaction which is smaller in absolute value than the main effect for Temperature 1 but larger than the main effect for Temperature 2. This interaction suggests that the effect of Temperature 1 depends at least somewhat on which level of Temperature 2 we use.

Step 4. Obtain the regression equation.

Choose **Stat**>**Regression**>**Regression.** Place Y in the Response: text box. Place X1, X2 and X1X2 in the Predictors: text box. Choose **OK.**

The Minitab Output

Regression Analysis

```
The regression equation is
Y = 2.66 - 0.0850 X1 + 0.0150 X2 + 0.0400 X1X2

Predictor        Coef        StDev            T          P
Constant      2.66000      0.01837       144.79      0.000
X1           -0.08500      0.01837        -4.63      0.010
X2            0.01500      0.01837         0.82      0.460
X1X2          0.04000      0.01837         2.18      0.095


S = 0.05196      R-Sq = 87.0%      R-Sq(adj) = 77.3%


Analysis of Variance
Source        DF         SS           MS          F         P
Regression     3     0.072400     0.024133       8.94     0.030
Error          4     0.010800     0.002700
Total          7     0.083200

Source        DF       Seq SS
X1             1     0.057800
X2             1     0.001800
X1X2           1     0.012800
```

Figure 7.1

The Minitab output, as shown in Figure 7.1, indicates the burst strength depends upon at least one of the factors since the p-value (0.030) is less than $\alpha = .10$. Further, we observe that $R^2 = 87.0\%$, which supports the idea that this is a reasonable model.

The Minitab output also indicates that the main effect of Temperature 1 is important since its p-value of 0.010 is much less than 0.10. In addition, the interaction between the two temperatures with a p-value of 0.095 also appears to be important.

Follow this step to construct a plot of the interaction.
Choose **Stat**>**ANOVA**>**Interactions Plot.** Place X2 and X1 in the
Factors: text box (order is important!). Darken the Raw response data in: option button. Place Y in the Source of response Raw response data in: text box.
Enter a title.
Type an appropriate title in the Title: text box. Choose **OK.**

The Minitab Output

Burst Strength of Packages

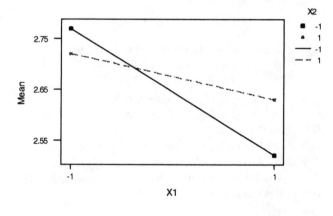

Figure 7.2

The Minitab output, as shown in Figure 7.2, indicates that the lines are not parallel and indicate the presence of an interaction. In this case, the burst strength appears to be at a maximum when Temperature 1 is at $220^0 F$ and Temperature 2 is at $120^0 F$. The burst strength appears to be a minimum when Temperature 1 is at $230^0 F$ and Temperature 2 is at $120^0 F$. For both levels of Temperature 2, the burst strength goes down as we increase Temperature 1. In this case, the interaction means that how much the burst strength goes down depends upon the specific level of Temperature 2 used. Finally, this plot indicates that otherwise, Temperature 2 has a minor effect on burst strength, which was reflected in the small size of the effect for this temperature. Observe that this Minitab analysis is consistent with the conceptual analysis.

Exercises

7.1 Buckner, Cammenga, and Weber (1993) ran a two factor experiment looking at the effects of pressure and the H_2/WF_6 ratio to determine their effect on the uniformity of the titanium nitride adhesion layer for a type of silicon wafer. We extracted the following 2^2 factorial design from their larger experiment. The levels for these two factors were

Factor	Low Level	High Level
Press.	15 Torr	70 Torr

The results follow.

Press.	H_2/WF_6	Uniformity
15	3	8.6
70	3	3.4
15	9	6.9
70	9	5.1

Follow these steps to estimate both main effects and the interaction. For this situation, the model is

$$y_i = \beta_0 + \beta_1 x_{i1} + \beta_2 x_{i2} + \beta_{12} x_{i1} x_{i2} + \epsilon_u$$

where

(A) y_i is the uniformity for the ith wafer tested,

(B) $x_{i1} = \begin{cases} -1 \text{ if pressure is 15 Torr} \\ 1 \text{ if Temp. 1 is 70 Torr} \end{cases}$,

(C) $x_{i2} = \begin{cases} -1 \text{ if } H_2/WF_6 \text{ ratio is 3} \\ 1 \text{ if } H_2/WF_6 \text{ ratio is 9} \end{cases}$,

(D) β_0 is the y-intercept,

(E) β_1 is the coefficient associated for pressure,

(F) β_2 is the coefficient associated for H_2/WF_6,

(G) β_{12} is the interaction coefficient for pressure and H_2/WF_6, and

(H) ϵ_i is a random error.

Step 1. Enter data.

Enter the pressure codes in column C1. Name column C1 as *X1*. Enter the H_2/WF_6 codes in column C2. Name column C2 as *X2*. Enter the "Uniformity" measurement (y_i's) in column C3. Name column C3 as *Y*.

Step 2. Introduce the interaction variable.

Choose **Calc**>**Calculator**. Place X1X2 in the <u>S</u>tore result in variable: text box. Place X1*X2 in the <u>E</u>xpression: text box. Choose **OK**.

Step 3. Estimate the effects.

(See the text for an explanation for these equations.) Choose **Calc**>**Calculator**. Type A in the Store result in variable: text box. Choose Statistics from the Functions: drop down dialog box. Select Sum from the Functions: list box. Place Sum(X1*Y)/4 in the Expression: text box. Choose **OK**.

Choose **Calc**>**Calculator**. Type B in the Store result in variable: text box. Place Sum(X2*Y)/4 in the Expression: text box. Choose **OK**.

Choose **Calc**>**Calculator**. Type AB in the Store result in variable: text box. Place Sum(X1X2*Y)/4 in the Expression: text box. Choose **OK**.

Use {Ctrl} D to go to the Data window.

Step 4. Obtain the regression equation.

Choose **Stat**>**Regression**>**Regression**. Place Y in the Response: text box. Place Xl, X2 and X1X2 in the Predictors: text box. Choose **OK**. Are the results consistent with the estimates?

Step 5. Construct the interaction plot.

Choose **Stat**>**ANOVA**>**Interactions Plot**. Place X2 and X1 in the Factors: text box (order is important!). Place Y in the Source of response Raw response data in: text box. Place an appropriate title in the Title: text box. Choose **OK**.

7.2 Said et al. (1994) studied the effect the calcination temperature and the mole fraction of cobalt on the surface area of an iron-cobalt hydroxide catalyst. From their work, we could extract the following 2^2 factorial experiment.

Temp.	*mole fraction*	*Surface Area*
200	0.6	90.6
600	0.6	25.0
200	2.8	40.9
600	2.8	19.0

Estimate both main effects and the interaction.

7.3 (EXO7O3) A chemical engineer used a replicated 22 factorial design to study the impact of inlet feed temperature and the reflux ratio upon the yield of gasoline from a distillation column. The yield in this case was defined to be the proportion of gasoline in the outlet stream from the top of the column. The experimental results follow.

	Reflux	
InletTemp.	*Ratio*	*Yield*
550	4	88
600	4	90
550	8	95
600	8	97
550	4	87
600	4	91
550	8	94
600	8	98

Estimate the two main effects and the interaction.

7.2   The 2^k Factorial Design

New Minitab Commands (and some Minitab commands used previously)

1. **Manip>Stack/Unstack>Stack** - Stacks a list of columns, one on top of the other, to form a new, longer column, and optionally stores a second column containing subscript (or group) values. In this section, you will use this command to stack data for determining important effects in a 2^k factorial design.

2. **Stat>Basic Statistics>Normality Test** - Generates a normal probability plot.

3. The grid on the graph resembles the grids found on normal probability paper. The vertical axis has a probability scale; the horizontal axis, a data scale. A least-squares line is fit to the plotted points and drawn on the plot for reference. The line forms an estimate of the cumulative distribution function for the population from which data are drawn. Numerical estimates of the population parameters, μ and σ, are displayed with the plot. In this section, you will obtain a normality probability to determine if the estimated effects (coefficients) follow normal distributions.

We can easily extend these techniques to k factors. We shall illustrate the extension through examples involving three factors.

Example 7.2 - Springs with Cracks
Box and Bisgaard (1987) discuss a manufacturing operation for carbon-springs with have a severe problem with cracks. Basic metallurgy suggests that the cracking depends upon

        (A) the temperature of the steel before quenching,
        (B) the amount of carbon in the formulation, and
        (C) the temperature of the quenching oil.

For this situation, the model is

$$y_i = \beta_0 + \beta_1 x_{i1} + \beta_2 x_{i2} + \beta_3 x_{i3} +$$
$$\beta_{12} x_{i1} x_{i2} + \beta_{13} x_{i1} x_{i3} + \beta_{23} x_{i2} x_{i3} +$$
$$\beta_{123} x_{i1} x_{i2} x_{i3} + \epsilon_u$$

where

(A) y_i is the per cent of springs which do not crack in the ith production lot,

(B) $x_{i1} = \begin{cases} \text{-1 if the steel temperature before quenching is } 1450^0 F \\ \text{1 if the steel temperature before quenching is } 1600^0 F \end{cases}$,

(C) $x_{i2} = \begin{cases} \text{-1 if the amount of carbon is 0.50\%} \\ \text{1 if the amount of carbon is 0.70\%} \end{cases}$,

(D) $x_{i3} = \begin{cases} \text{-1 if the temperature of the quenching oil is } 70^0 F \\ \text{1 if the temperature of the quenching oil is } 120^0 F \end{cases}$,

(E) β_0 is the y-intercept,

(F) β_1, β_2, and β_3 are the coefficients associated with x_{i1}, x_{i2}, and x,

(G) β_{12}, β_{13}, and β_{23} are the coefficients for the two-factor interactions,

(H) β_{123} is the coefficient for the three-factor interaction, and

(I) ϵ_i is a random error.

The experimental results follow.

x_1	x_2	x_3	y
-1	-1	-1	67
1	-1	-1	79
-1	1	-1	61
1	1	-1	75
-1	-1	1	59
1	-1	1	90
-1	1	1	52
1	1	1	87

Follow these four steps to estimate the effects and to determine the least squares multiple regression equation between the variables.

Step 1. Enter data.

Enter the codes for the steel temperature before quenching in column C1. Name column C1 as *X1*. Enter the codes for the amount of carbon in column C2. Name column C2 as *X2*. Enter the codes for the temperature of the quenching oil in column C3. Name column C3 as *X3*. Enter the percent of springs which do not crack($y_i's$) in column C4. Name column C4 as *Y*.

Step 2. Introduce the interaction variables.

Choose **Calc**>**Calculator**. Type *X1X2* in the Store result in variable: text box. Place X1*X2 in the Expression: text box. Choose **OK**.

Choose **Calc**>**Calculator**. Type *X1X3* in the Store result in variable: text box. Place X1*X3 in the Expression: text box. Choose **OK**.

Choose **Calc**>**Calculator**. Type *X2X3* in the Store result in variable: text box. Place X2*X3 in the Expression: text box. Choose **OK**.

Choose **Calc**>**Calculator**. Type *X1X2X3* in the Store result in variable: text box. Place X1*X2*X3 in the Expression: text box. Choose **OK**.

Step 3. Estimate the effects.

(See the text for an explanation for these equations.) Choose **Calc**>**Calculator**. Type *A* in the Store result in variable: text box. Choose Statistics from the Functions: drop down dialog box. Select Sum from the Functions: list box. Place Sum(X1*Y)/4 in the Expression: text box. Choose **OK**.

Choose **Calc**>**Calculator**. Type *B* in the Store result in variable: text box. Place Sum(X2*Y)/4 in the Expression: text box. Choose **OK**.

Choose **Calc**>**Calculator**. Type *C* in the Store result in variable: text box. Place Sum(X3*Y)/4 in the Expression: text box. Choose **OK**.

Choose **Calc**>**Calculator**. Type *AB* in the Store result in variable: text box. Place Sum(X1X2*Y)/4 in the Expression: text box. Choose **OK**.

Choose **Calc**>**Calculator**. Type *AC* in the Store result in variable: text box. Place Sum(X1X3*Y)/4 in the Expression: text box. Choose **OK**.

Choose **Calc**>**Calculator**. Type *BC* in the Store result in variable: text box. Place Sum(X2X3*Y)/4 in the Expression: text box. Choose **OK**.

Choose **Calc**>**Calculator**. Type *ABC* in the Store result in variable: text box. Place Sum(X1X2X3*Y)/4 in the Expression: text box. Choose **OK**.

Use {Ctrl} D to go to the Data window. Under A we see that the estimate of the main effect of the steel temperature (A) is 23.0, the main effect of the amount of carbon is -5, and the main effect of the temperature of the quenching oil is 1.5. Similarly, the estimate of the effect of the interaction of steel temperature and carbon (AB) is 1.5, the estimate of the effect of the interaction of steel temperature and oil temperature (AC) is 10, and estimate of the effect of the interaction of carbon and oil is 0.0. The estimate of the three factor interaction (ABC) is 0.5.

Step 4. Obtain the regression equation.

Choose **Stat**>**Regression**>**Regression**. Place Y in the Response: text box. Place Xl, X2, X3,X1X2, X1X3,X2X3 and X1X2X3 in the Predictors: text box. Choose **OK**.

The Minitab Output

Regression Analysis

```
The regression equation is
Y = 71.2 + 11.5 X1 - 2.50 X2 + 0.750 X3 + 0.750 X1X2 + 5.00 X1X3
       +0.000000 X2X3 + 0.250 X1X2X3
```

Predictor	Coef	StDev	T	P
Constant	71.2500	0.0000	*	*
X1	11.5000	0.0000	*	*
X2	-2.50000	0.00000	*	*
X3	0.750000	0.000000	*	*
X1X2	0.750000	0.000000	*	*
X1X3	5.00000	0.00000	*	*
X2X3	0.00000000	0.00000000	*	*
X1X2X3	0.250000	0.000000	*	*

```
S = *
```

Analysis of Variance

Source	DF	SS	MS	F	P
Regression	7	1317.500	188.214	*	*
Error	0	*	*		
Total	7	1317.500			

Source	DF	Seq SS
X1	1	1058.000
X2	1	50.000
X3	1	4.500
X1X2	1	4.500
X1X3	1	200.000
X2X3	1	0.000

Figure 7.3

The Minitab output, as shown in Figure 7.3, when we estimate the full model is given by

$$y_i = \beta_0 + \beta_1 x_{i1} + \beta_2 x_{i2} + \beta_3 x_{i3} +$$
$$\beta_{12} x_{i1} x_{i2} + \beta_{13} x_{i1} x_{i3} + \beta_{23} x_{i2} x_{i3} +$$
$$\beta_{123} x_{i1} x_{i2} x_{i3} + \epsilon_u$$

Because our model has as many terms as we have observations, we are unable to estimate an error term. As a result, we are unable to perform any formal tests, which is a problem.

Using Normal Probability Plots

The normal probability plot provides a graphical method for determining important effects in a 2^k factorial design. If the assumptions for our regression model are reasonable, then all of the estimated effects (or estimated coefficients) follow normal distributions. With factorial experiments, if all of the true effects (or coefficients) are zero, then a normal probability plot should form a straight line.

Follow these steps to construct a normal probability plot.

Step 1. Stack the estimates.
   Choose **Manip**>**Stack.** Place A, B, C, AB, AC, BC, and ABC in the Stack the following columns: text box. Place Effects in the Store the stacked data in: text box. Choose **OK.**
Step 2. Construct the normal probability plot.
   Choose **Stat**>**Basic Statistics**>**Normality Test.** Place Effects in the Variable: text box. Place an appropriate title in the Title: text box. Choose **OK.**

The Minitab Output

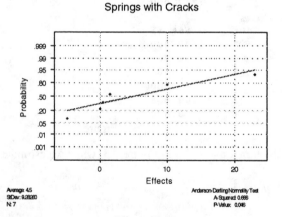

Figure 7.4

The Minitab output, as shown in Figure 7.4, indicates that the steel temperature (A) and the interaction of steel temperature and oil quench temperature (AC) appear most important since they appear to be outliers and do not lie on the straight line defined by the estimated effects smallest in absolute value.

Follow this step to construct a plot of the interaction.
Choose **Stat**>**ANOVA**>**Interactions Plot.** Place X3 and X1 in the Factors: text box (order is important!). Darken the Raw response data in: option button. Place Y in the Raw response data in: text box. Place an appropriate title in the Title: text box. Choose **OK.**

The Minitab Output

Interaction Plot - Means for Y

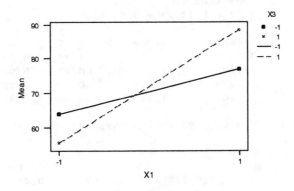

Figure 7.5

This Minitab output, as shown in Figure 7.5, indicates that the per cent of springs without cracks increases as the steel temperature increases (x_1) increases, which makes a great deal of engineering sense. In this case, the per cent of springs without cracks appears to be at a maximum when the temperature of the quenching oil is $120^0 F(x_3 = 1)$ and the steel temperature before quenching is $1600^0 F(x_1 = 1)$. The per cent of springs without cracks appears to be at a minimum when the temperature of the quenching oil is $120^0 F(x_3 = 1)$ and the steel temperature before quenching is $1450^0 F(x_1 = -1)$.

Since the estimated main effect of the amount of carbon (B) and all interactions involving carbon are small, *at least over the range of carbon studied*, this factor will be dropped from the analysis.

Follow this step to determine the least squares multiple regression equation between the variables.

Choose **Stat**>**Regression**>**Regression.** Place Y in the

Response: text box. Place Xl, X3, and X1X3 in the Predictors: text box. Choose **OK.**

The Minitab Output

Regression Analysis

```
The regression equation is
Y = 71.2 + 11.5 X1 + 0.75 X3 + 5.00 X1X3
```

Predictor	Coef	StDev	T	P
Constant	71.250	1.311	54.35	0.000
X1	11.500	1.311	8.77	0.001
X3	0.750	1.311	0.57	0.598
X1X3	5.000	1.311	3.81	0.019

```
S = 3.708        R-Sq = 95.8%       R-Sq(adj) = 92.7%
```

Analysis of Variance

Source	DF	SS	MS	F	P
Regression	3	1262.50	420.83	30.61	0.003
Error	4	55.00	13.75		
Total	7	1317.50			

Source	DF	Seq SS
X1	1	1058.00
X3	1	4.50
X1X3	1	200.00

Figure 7.6

The overall F test, as shown in Figure 7.6, has a p-value of 0.003, indicating that the cracking depends upon at least one of the factor. The R^2 indicates that the model explains over 95% of the total variability. The tests on the individual factors confirm that the steel temperature main effect and the steel temperature by oil quench temperature interaction.

Exercises

7.4 Consider the polymer filament example where the breaking strength of the filament depends upon the amount of the catalyst (Factor A), the polymerization temperature (Factor B), and the polymerization pressure (Factor C). Suppose the engineers assigned to this process need to conduct an appropriate two-level full factorial design. They have chosen for the levels the following.

Catalyst	Temp.	Press.
1.5%	$250^0 C$	25 psig
3.0%	$280^0 C$	40 psig

a. Give the appropriate factorial design in the design variables.

b. Give the appropriate design in the natural units.

c. Give the appropriate design in random order.

d. Repeat the above adding 4 center runs to the design.

7.5 Eibl, Kess, and Pukelsheim (1992) used a 2^3 factorial design to study the impact of belt speed (Factor A), tube width (Factor B), and pump pressure (Factor C) on the coating thickness for a painting operation. The responses follow.

(1)	0.575
a	0.585
b	0.680
ab	0.590
c	0.665
ac	0.585
bc	0.915
abc	0.785

a. Estimate all of the main effects and interactions. Use software as appropriate.

b. Plot the estimated effects on a normal probability plot. Again use software as appropriate.

c. Interpret your results.

7.6 (EXO711) Ferrer and Romero (1995) conducted an unreplicated 2^4 factorial design to determine the effects of the amount of glue (A), predrying temperature (B), tunnel temperature (C), and pressure (D) on the adhesive force obtained in an adhesive process of polyurethane sheets. The results follow.

A	B	C	D	y
-1	-1	-1	-1	3.80
1	-1	-1	-1	4.34
-1	1	-1	-1	3.54
1	1	-1	-1	4.59
-1	-1	1	-1	3.95
1	-1	1	-1	4.83
-1	1	1	-1	4.86
1	1	1	-1	5.28
-1	-1	-1	1	3.29
1	-1	-1	1	2.82
-1	1	-1	1	4.59
1	1	-1	1	4.68
-1	-1	1	1	2.73
1	-1	1	1	4.31
-1	1	1	1	5.16
1	1	1	1	6.06

Follow these steps to estimate the effects and to determine the least squares multiple regression equation between the variables.

Step 1. Enter data.

Enter the codes for the amount of glue (A) in column C1. Name column C1 as *X1*. Enter the codes for the predrying temperature (B) in column C2. Name column C2 as *X2*. Enter the codes for the tunnel temperature (C) in column C3. Name column C3 as *X3*. Enter the codes for the pressure (D) in column C4. Name column C4 as *X4*. Enter the adhesive forces ($y_i's$) in column C5. Name column C5 as *Y*.

Step 2. Introduce the interaction variables.

Choose **Calc**>**Calculator**. Type *X1X2* in the **S**tore result in variable: text box. Place X1*X2 in the **E**xpression: text box. Choose **OK**.
Choose **Calc**>**Calculator**. Type *X1X3* in the **S**tore result in variable: text box. Place X1*X3 in the **E**xpression: text box. Choose **OK**.
Choose **Calc**>**Calculator**. Type *X1X4* in the **S**tore result in variable: text box. Place X1*X4 in the **E**xpression: text box. Choose **OK**.
Choose **Calc**>**Calculator**. Type *X2X3* in the **S**tore result in variable: text box. Place X2*X3 in the **E**xpression: text box. Choose **OK**.
Choose **Calc**>**Calculator**. Type *X2X4* in the **S**tore result in variable: text box. Place X2*X4 in the **E**xpression: text box. Choose **OK**.
Choose **Calc**>**Calculator**. Type *X3X4* in the **S**tore result in variable: text box. Place X3*X4 in the **E**xpression: text box. Choose **OK**.

Choose **Calc**>**Calculator**. Type *X1X2X3* in the **S**tore result in variable: text box. Place X1*X2*X3 in the **E**xpression: text box. Choose **OK**.
Choose **Calc**>**Calculator**. Type *X1X2X4* in the **S**tore result in variable: text box. Place X1*X2*X4 in the **E**xpression: text box. Choose **OK**.
Choose **Calc**>**Calculator**. Type *X2X3X4* in the **S**tore result in variable: text box. Place X2*X3*X4 in the **E**xpression: text box. Choose **OK**.

Choose **Calc**>**Calculator**. Type *X1X2X3X4* in the **S**tore result in variable: text box. Place X1*X2*X3*X4 in the **E**xpression: text box. Choose **OK**.

Step 3. Estimate the effects.

(See the text for an explanation for these equations.) Choose **Calc**>**Calculator**. Type *A* in the **S**tore result in variable: text box. Place Sum(X1*Y)/8 in the **E**xpression: text box. Choose **OK**.
Choose **Calc**>**Calculator**. Type *B* in the **S**tore result in variable: text box. Place Sum(X2*Y)/8 in the **E**xpression: text box. Choose **OK**.
Choose **Calc**>**Calculator**. Type *C* in the **S**tore result in variable: text box. Place Sum(X3*Y)/8 in the **E**xpression: text box. Choose **OK**.
Choose **Calc**>**Calculator**. Type *D* in the **S**tore result in variable: text box. Place Sum(X4*Y)/8 in the **E**xpression: text box. Choose **OK**.

Choose **Calc**>**Calculator**. Type *AB* in the **S**tore result in variable: text box. Place Sum(X1X2*Y)/8 in the **E**xpression: text box. Choose **OK**.
Choose **Calc**>**Calculator**. Type *AC* in the **S**tore result in variable: text box. Place Sum(X1X3*Y)/8 in the **E**xpression: text box. Choose

OK.

Choose **Calc**>**Calculator**. Type *AD* in the Store result in variable: text box. Place Sum(X1X4*Y)/8 in the Expression: text box. Choose **OK.**

Choose **Calc**>**Calculator**. Type *BC* in the Store result in variable: text box. Place Sum(X2X3*Y)/8 in the Expression: text box. Choose **OK.**

Choose **Calc**>**Calculator**. Type *BD* in the Store result in variable: text box. Place Sum(X2X4*Y)/8 in the Expression: text box. Choose **OK.**

Choose **Calc**>**Calculator**. Type *CD* in the Store result in variable: text box. Place Sum(X3X4*Y)/8 in the Expression: text box. Choose **OK.**

Choose **Calc**>**Calculator**. Type *ABC* in the Store result in variable: text box. Place Sum(X1X2X3*Y)/8 in the Expression: text box. Choose **OK.**

Choose **Calc**>**Calculator**. Type *ABD* in the Store result in variable: text box. Place Sum(X1X2X4*Y)/8 in the Expression: text box. Choose **OK.**

Choose **Calc**>**Calculator**. Type *BCD* in the Store result in variable: text box. Place Sum(X2X3X4*Y)/8 in the Expression: text box. Choose **OK.**

Choose **Calc**>**Calculator**. Type *ABCD* in the Store result in variable: text box. Place Sum(X1X2X3X4*Y)/8 in the Expression: text box. Choose **OK.**

Use {Ctrl} D to go to the Data window. Record the estimates all of the main effects and interactions.

2. Plot the estimated effects on a normal probability plot. Again use software as appropriate.

3. Interpret your results.

7.3   Half Fractions of the 2^k Factorial

New Minitab Commands

1. **Stat**>**DOE**>**Create Factorial Design** - Generates two-level designs, either full or fractional factorials, and Plackett-Burman designs. In this section, you will use this command to create the negative half fraction factorial design in Example 7.3 - Oil Extraction from Peanuts.

2. **Stat**>**DOE**>**Analyze Factorial Design** - Fits two-level full and fractional factorial designs, and Plackett-Burman designs. In this section, you will use this

command to obtain estimate of the main and interaction effect for Example 7.3 - Oil Extraction from Peanuts.

3. **Stat>DOE>Factorial Plots** - Allows you to display main effects, interactions, and cube plots for designs generated with Create Factorial Design. You must fit a design using Analyze Factorial Design before you can display the factorial plots. In this section, you will use this command to obtain plots for the main effects in Example 7.3 - Oil Extraction from Peanuts.

For even moderate values of k, the 2^k factorial design can require a prohibitive number of experimental runs. Fractional factorial designs reduce the total number of treatment combinations while preserving the basic factorial structure of the experiment. The "main effects" principle states that main effects tend to dominate two-factor interactions, while two-factor interactions tend to dominate three-factor interactions, etc. The full 2^k factorial design allows us to estimate all of the interactions; as k gets larger, more and more of these higher order interactions probably not important. Fractional factorial designs "sacrifice" the ability to estimate the higher order interactions in order to reduce the number of treatment combinations.

Half Fraction of the 2^3

Recall the steel springs with cracks example where we wish to determine the influence of steel temperature (Factor A), amount of carbon (Factor B), and oil quench temperature (Factor C) on the per cent of steel springs which exhibit no signs of cracking. Under the main effects principle, the one effect least likely to be important is the three-factor interaction. We need a design with only four treatment combinations. The following table gives the resulting design.

x_1	x_2	x_3
1	-1	-1
-1	1	-1
-1	-1	1
1	1	1

The resulting table of contrasts

I	x_1	x_2	x_3	x_1x_2	x_1x_3	x_2x_3	$x_1x_2x_3$
1	1	-1	-1	-1	-1	+1	+1
1	-1	1	-1	-1	+1	-1	+1
1	-1	-1	1	+1	-1	-1	+1
1	1	1	1	+1	+1	+1	+1

indicates that the column for the three-factor interaction ($x_1x_2x_3$) is the same as the intercept (I) column. As a result, the numerator for estimating the intercept is exactly the same numerator for estimating the three-factor interaction ($x_1x_2x_3$). Thus, neither can be uniquely estimated! In this situation, the tree-factor interaction, ABC, is said to be **aliased** with the intercept. Observe that all of the main effects are aliased with the two-factor interactions.

If we use $x_1x_2x_3 = 1$ to select the treatment combinations, the resulting design is called the positive half fraction of the 2^3. If we use $x_1x_2x_3 = -1$ to select the treatment combinations, the resulting design is called the negative half fraction of the 2^3.

Example 7.3 - Oil Extraction from Peanuts

Kilgo (1988) performed an experiment to determine the effect of CO_2 pressure, CO_2 temperature, peanut moisture, CO_2 flow rate, and peanut particle size on the total yield of oil per batch of peanuts. The specific levels used follow.

A	B	C	D	E
Press.	Temp.	Moist.	Flow	Part. Size
bar	deg. C	% by weight	liters/min	mm
415	25	5	40	1.28
550	95	15	60	4.05

For economic reasons, a half fraction of the 2^5, a 2^{5-1} design from the 2^4 full factorial design was selected. By letting $x_5 = -x_1x_2x_3x_4$, we obtain the negative half fraction with ABCDE as the defining interaction (We will use this when we construct a main effects plot). The experimental results follow.

x_1	x_2	x_3	x_4	Re *sponse*	*Yield*
-1	-1	-1	-1	1	63
1	-1	-1	-1	-1	21
-1	1	-1	-1	-1	36
1	1	-1	-1	1	99
-1	-1	1	-1	-1	24
1	-1	1	-1	1	66
-1	1	1	-1	1	71
1	1	1	-1	-1	54
-1	-1	-1	1	-1	23
1	-1	-1	1	1	74
-1	1	-1	1	1	80
1	1	-1	1	-1	33
-1	-1	1	1	1	63
1	-1	1	1	-1	21
-1	1	1	1	-1	44
1	1	1	1	1	96

Follow this step to determine the alias structure.

Choose **Stat>DOE>Create Factorial Design.** Darken the 2-level factorial (default generators) Type of design option button. Select 5 from the Number of factors: drop down list box.

Select a half fraction design.

Choose **Designs.** Select 1/2 fraction from the list box at the top of the Design subdialog box. Choose **OK.**

dialog box. Choose **OK.**

Select the negative half fraction.

Choose Options. Darken the Use fraction number: option button. Place 1 in the Use fraction number: text box (to obtain the negative half fraction with ABCDE as the defining interaction).

Select a standard order for the design.

Remove the check from the Randomize runs checkbox, producing a standard order for the design. Choose **OK.**

The Minitab Output

Factorial Design
Fractional Factorial Design

Factors:	5	Base Design:	5, 16	Resolution: V
Runs:	16	Replicates:	1	Fraction: 1/2, number 1
Blocks: none		Center pts (total):	0	

Design Generators: E = -ABCD

Alias Structure

I - ABCDE

A - BCDE
B - ACDE
C - ABDE
D - ABCE
E - ABCD
AB - CDE

AC - BDE
AD - BCE
AE - BCD
BC - ADE
BD - ACE
BE - ACD
CD - ABE
CE - ABD
DE - ABC

Figure 7.7

The Minitab output, as shown in Figure 7.7, indicates the alias structure. This alias structure tells us that the model can be estimated with terms for the intercept, each main effect, and each two-factor interaction.

Follow these steps to estimate the effects.

Step 1. Enter data.

Enter the yield of oil per batch of peanuts ($y_i's$) in column C9. Name column C9 as *Y*.

Step 2. Estimate the effects.

Choose **Stat>DOE>Analyze Factorial Design.** Place Y in the Responses: text box. Choose Terms. Place 2 in the Include terms in the model up through order: drop down list box. Place A:A, B:B, C:C, D:D, E:E, AB, AC, AD, AE, BC, BD, BE, CD, CE, and DE in the Selected terms: text box. Choose **OK.**

Store the results.

Choose Storage. Place a check in the Model Information Effects checkbox. Choose **OK.**

Obtain a normal probability plot of the effects.

Choose Graphs. Place a check in the Normal Effects plot checkbox. Choose **OK.**

The Minitab Output

Fractional Factorial Fit

Estimated Effects and Coefficients for Y

Term	Effect	Coef
Constant		54.25
A	7.50	3.75
B	19.75	9.87
C	1.25	0.62
D	0.00	0.00
E	-44.50	-22.25
A*B	5.25	2.62
A*C	1.25	0.62
A*D	-4.00	-2.00
A*E	-7.00	-3.50
B*C	3.00	1.50
B*D	-1.75	-0.88
B*E	-0.25	-0.13
C*D	2.25	1.12
C*E	6.25	3.12
D*E	-3.50	-1.75

Figure 7.8

The Minitab output, as shown in Figure 7.8, provides the estimates of the main and interaction effects. With factorial experiments, if all of the true effects (or coefficients) are zero, then a normal probability plot should form a straight line.

The Minitab Output

Normal Probability Plot of the Effects
(response is Y, Alpha = .10)

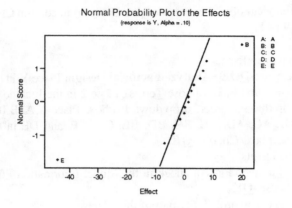

Figure 7.9

This plot, as shown in Figure 7.9, indicates that the two important effects are the main effect due to the average particle size (E, estimate 44.5) and the main effect due to temperature (B, estimate 19.75). They appear most important since they appear to be outliers and do not lie on the straight line defined by the estimated effects smallest in absolute value.

Follow this step to construct a plot of the main effects.
Choose **Stat>DOE>Factorial Plots.** Darken the Main effects (response versus levels of one factor) option button. Choose Setup. Place Y in the Responses: text box. Place B and E Selected: text box.
Enter a title.
Choose Options. Place an appropriate title in the Title: text box. Choose **OK.**
Choose **OK.**

The Minitab Output

Main Effects for Factors B and E

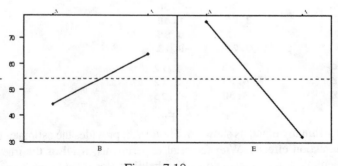

Figure 7.10

The plots, as shown in Figure 7.10, indicate that the yield decreases as the particle size (E) increases, which makes engineering sense. Similarly, the more energetic the physical system, the more likely transfer from the peanuts to the CO_2. Thus, the higher the temperature (B), the better the yield.

7.4 Smaller Fractions

New Minitab Commands

1. **Stat>DOE>Analyze Custom Design** - Allows you to fit designs that were not generated using the Minitab menu commands. You may enter the design data directly into the Data Window, or you can import the design from a data file. You can also analyze designs created with earlier versions of Minitab with the Analyze Custom Design command. In this section, you will use this command to analyze a fractional factorial design in Example 7.4 - Thickness of Paint Coatings. You will also use this command to obtain plots of the main effects.

For more than four factor ($k > 4$), we can construct even smaller fractions than the half.

Example 7.4 - Thickness of Paint Coatings

Eibl, Kess, and Pukelsheim (1992) used a sequential experimental strategy to achieve a target coating thickness of 0.8 mm for a painting process. In their first experiment, they ran a factorial design to study the impact of

- (A) belt speed (Factor A),
- (B) tube width (Factor B),
- (C) pump pressure (Factor C),
- (D) paint viscosity (Factor D),
- (E) tube height (Factor E), and
- (F) heating temperature (Factor F)

on this thickness. The experimental results follow.

x_1	x_2	x_3	x_4	x_5	x_6	Response	y
-1	-1	-1	1	-1	1	d	1.490
1	-1	-1	1	1	-1	ade	0.835
-1	1	-1	-1	1	-1	be	1.738
1	1	-1	-1	-1	1	abf	1.130
-1	-1	1	-1	1	1	cef	1.545
1	-1	1	-1	-1	-1	ac	0.980
-1	1	1	1	-1	-1	bcd	2.175
1	1	1	1	1	1	$abcdef$	1.448

Follow these steps to construct this 2^{6-3} factorial design.

Step 1. Enter the design matrix.

Enter the levels of factor A in column C1. Name column C1 as A. Enter the levels of factor B in column C2. Name column C2 as B. Enter the levels of factor C in column C3. Name column C3 as B.

Choose **C̲alc**>**Cal̲culator**. Place D in the S̲tore result in variable: text box. Place B*C in the E̲xpression: text box. Choose **OK**.
Choose **C̲alc**>**Cal̲culator**. Place E in the S̲tore result in variable: text box. Place A*B*C in the E̲xpression: text box. Choose **OK**.
Choose **C̲alc**>**Cal̲culator**. Place F in the S̲tore result in variable: text box. Place A*B in the E̲xpression: text box. Choose **OK**.

Step 2. Enter data.

Enter the yield of oil per batch of peanuts ($y_i's$) in column C7. Name column C7 as Y.

Step 3. Estimate the effects.

Choose **S̲tat**>**D̲OE**>**Analyze C̲ustom Design.** Darken the 2-L̲evel factorial Design option button.
Select the model.
Choose F̲it Model. Place Y in the R̲esponses: text box. Place A-F in the F̲actors: text box.
Obtain a normal probability plot of the effects.
Choose G̲raphs. Place a check in the N̲ormal Effects checkbox. Choose **OK**.
Display the effects in the Session window.
Place a check in the D̲isplay effects plot checkbox. Choose **OK**.

The Minitab Output

Fractional Factorial Fit

Estimated Effects and Coefficients for Y

Term	Effect	Coef
Constant		1.4176
A	-0.6388	-0.3194
B	0.4102	0.2051
C	0.2387	0.1194
D	0.1388	0.0694
E	-0.0522	-0.0261
F	-0.0287	-0.0144
A*C	-0.0072	-0.0036

Figure 7.11

The Minitab output, as shown in Figure 7.11, gives the estimated effects, and Figure 7.12 is the resulting normal probability plot. In this case the normal probability plot suggests that the estimated effect associated with the belt speed (A) appears to be important. The estimated effect for tube width (B) may be important because it does not appear to fall on the line generated by the effects near zero.

The Minitab Output

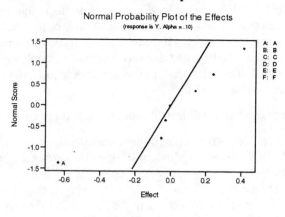

Figure 7.12

We can model the response variable (coating thickness) from belt speed (A), tube width (B), and their interaction (AB) can provide additional insights for the experiment.

Follow these steps to obtain the regression equation.
Choose **S**tat>**Regression**>**Regression**. Place Y in the Response: text box. Place A, B and F in the Predictors: text box. Choose **OK**.

The Minitab Output

Regression Analysis

```
The regression equation is
Y = 1.42 - 0.319 A + 0.205 B - 0.0144 F
```

Predictor	Coef	StDev	T	P
Constant	1.41762	0.07028	20.17	0.000
A	-0.31937	0.07028	-4.54	0.010
B	0.20512	0.07028	2.92	0.043
F	-0.01438	0.07028	-0.20	0.848

```
S = 0.1988     R-Sq = 88.0%     R-Sq(adj) = 78.9%
```

Analysis of Variance

Source	DF	SS	MS	F	P
Regression	3	1.15427	0.38476	9.74	0.026
Error	4	0.15807	0.03952		
Total	7	1.31234			

Figure 7.13

The Minitab output, as shown in Figure 7.13, indicates, from the overall F-test, with a p-value of 0.026, that at least one of the terms is important. The R^2 of 88.0% indicates that the model explains almost 90% of the total variability. The test on the estimated coefficients indicate that belt speed (A) is reasonably important, with a p-value of 0.010, and that tube width (B) is probably important, with a p-value of 0.043. The belt speed -tube width interaction (AB) is clearly not important, with a p-value of 0.848.

A plot of the main effects will provide addition support in our analysis.

Follow this step to construct a plot of the main effects.
Choose **Stat**>**DOE**>**Analyze Custom Design.** Darken the 2-Level factorial Design option button.
Setup the plot of the main effects.
Choose Plot Response. Place a check in the Main effects (response versus levels of 1 factor) checkbox. Choose Setup. Place Y in the Responses: text box. Place A and B in the Selected: text box. Choose **OK.**
Choose **OK.**

The Minitab Output

Main Effects for Y

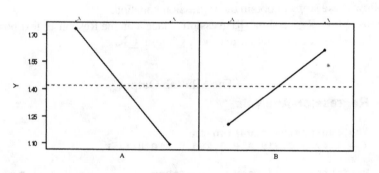

Figure 7.14

The Minitab output, as shown in Figure 7.14, shows the main effects plots for both factors. The main effects plot for the belt speed (A, x_1) shows that the higher the belt speed, the thinner the coating. The main effects plot for the tube width (B, x_2) shows that the thickness increases as we increase the tube width. Taken together, these plots suggest using the high level of paint speed and the low level of tube width to reduce the thickness of the coating.

7.5   Design Resolution

Design resolution refers to the length (number of letters) in the smallest defining or "generalized" interaction and tells the user some critical information about the alias structure. A Resolution III design has a defining interaction of three letters.

A Resolution IV design has a defining interaction of four letters and a Resolution V design has a defining interaction of five letters.

Exercises

7.7 Aceves-Mijares and colleagues (1995) consider a polysilicon deposition process commonly used in the manufacture of integrated circuits. Four factors of interest are pressure (A), temperature gradient (B), furnace temperature (C), and the location of the gas input (D). Suppose the engineers assigned to this process wish to conduct and eight-run design using the following levels.

Pressure	Temp. Gradient	Furnace Temp.	Gas Input
1 Torr	Flat	625°C	Bottom
3 Torr	Ramp	725°C	Top

a. Construct the appropriate fractional factorial design, in the design variables, that has the highest resolution possible.

b. Give this design in natural units.

c. Give this design in random order.

7.8 Shoemaker, Tsui, and Wu (1991) study the process which grows silicon wafers for integrated circuits. These layers need to be as uniform as possible since later processing steps form electrical devices within these layers. The smallest unit to which the engineers can apply the processing steps is a batch of wafers. The individual wafers are the observational units. In their initial experiment, the engineers seek to determine which of six possible factors truly influence the uniformity of the silicon layer. They intend to conduct follow up experimentation using the significant factors from this initial experiment. To minimize the size of the design, they use only two levels for each factor. The factors and their levels follow.

Deposition Temp.	1210	1220
Deposition Time	low	high
Argon Flow Rate	55%	59%
HCl Etch Temp.	1180	1215
HCl Flow Rate	10%	14%
Nozzle Position	2	4

Construct a 16 run Resolution IV (2_{IV}^{6-2}) fractional factorial design in both the design variables and the natural units.

7.9 (EXO715) Anand, Bhadkamkar, and Moghe (1995) used a fractional factorial design to determine which of five possible factors influenced the determination of carbon in cast iron. The five factors and their levels follow.

		Morning	Afternoon
A - Testing Time		Morning	Afternoon
B - KOH Conc.		38	42
C - Heating Time		45	75
D - Oxygen Flow		Slow	Fast
E - Muffle Temp.		950	1100

The actual experiment and its results follow.

-1	-1	-1	-1	-1	3.130
-1	-1	1	1	-1	3.065
-1	1	-1	-1	1	3.105
-1	1	1	1	1	2.940
1	-1	-1	1	1	2.940
1	-1	1	-1	1	3.110
1	1	-1	1	-1	3.145
1	1	1	-1	-1	3.240

    a. Identify this design.

    b. Give the defining interaction or interactions.

    c. Analyze the experimental results.

7.10 (EXO716) Pignatiello and Ramberg (1985) studied the impact of several factors involving the heat treatment of leaf springs. In this process, a conveyor system transports leaf spring assemblies through a high temperature furnace. After this heat treatment, a high pressure press induces the curvature. After leaving the press, an oil quench cools the spring to near ambient temperature. An important quality characteristic of this process is the resulting free height of the spring. They used the following factors.

Factor	Low Level	High Level
High Heat Temp. (x_1)	1840	1880
Heating Time (x_2)	23	25
Transfer Time (x_3)	10	12
Hold Down Time (x_4)	2	3

The experimental results follow.

x_1	x_2	x_3	x_4	y
-1	-1	-1	-1	7.37
1	-1	-1	1	7.66
-1	1	-1	1	7.67
1	1	-1	-1	7.79
-1	-1	1	1	7.52
1	-1	1	-1	7.64
-1	1	1	-1	7.54
1	1	1	1	7.90

Follow these steps to help you identify the design and the defining interaction(s).

Step 1. Obtain the design matrix.

Choose **Stat>DOE>Create Factorial Design.** Darken the 2-level factorial (default generators) Type of design option button. Select 4 from the Number of factors: drop down list box.

Select the half fraction design.

Choose Designs. Select 1/2 fraction from the list box at the top of the Design subdialog box. Choose **OK.**

Select the factors.

Choose Factors. Place 0 in the Low text boxes for factors A, B, C and D. Choose **OK.**

Obtain a standard order for the design.

Choose Options. Remove the check from the Randomize runs checkbox, producing a standard order for the design. Choose **OK.**

Choose **OK.**

Look at the Session window to help you identify this design and the defining interactions(s).

Step 2. Enter data.

Enter the free height of the springs ($y_i's$) in column C8. Name column C8 as Y.

Step 3. Estimate the effects.

Choose **Stat>DOE>Analyze Factorial Design.** Place Y in the Responses: text box. Choose Terms. Place 2 in the Include terms in the model up through order: drop down list box. Place A:A, B:B, C:C, D:D, AB, AC and AD in the Selected terms: text box. Choose **OK.**

Store the design information.

Choose Storage. Place a check in the Model Information Effects checkbox. Choose **OK.** Choose Graphs.

Obtain a normal probability plot of the effects.

Place a check in the Normal Effects plot checkbox. Choose **OK.** Analyze the experimental results.

7.11 (EXO717) Anand, Bhadkamkar, and Moghe (1995) used a fractional factorial design to determine which of six possible factors influenced the determination of manganese in cast iron. The six factors and their levels follow.

	Medium	Fast
A - Titration Speed		
B - Dissolution Time	20	30
C - $AgNO_3$ Addition	20	10
D - Persulfate Addition	2	3
E - Vol. $HMnO_4$	100	150
F - Sodium Arsenite	0.10	0.15

The actual experiment and its results follow.

-1	-1	-1	-1	-1	-1	0.1745
-1	-1	-1	1	1	1	0.1340
-1	1	1	-1	-1	1	0.1630
-1	1	1	1	1	-1	0.0680
1	-1	1	-1	1	-1	0.0055
1	-1	1	1	-1	1	0.0330
1	1	-1	-1	1	1	0.0690
1	1	-1	1	-1	-1	0.0730

1. Identify this design.
2. Give the defining interaction or interactions.
3. Analyze the experimental results.

7.6   Central Composite Designs

New Minitab Commands (and some Minitab commands used previously)

1. **Stat>DOE>Create RS Design** - Generates Box-Behnken and central composite response surface designs. In this section, you will use this command to create the central composite design in Example 7.5 - Optimizing the Deposition Rate for a Silicon Wafer.

2. **Stat>DOE>Analyze RS Design** - Fits response surface models generated with Create RS Design. In this section, you will analyze the model in Example 7.5 - Optimizing the Deposition Rate for a Silicon Wafer.

3. **Stat>DOE>RS Plots** - Displays contour, wireframe and surface plots. In this section, you will use this command to obtain a surface plot in Example 7.5 - Optimizing the Deposition Rate for a Silicon Wafer.

Engineering experiments, particularly if we seek to optimize a process or product, should proceed sequentially. The most commonly used design to estimate a model that has at least as many distinct treatment combinations as terms in the models and has at least three levels for each factor is the central composite design.

A Central Composite Design
Example 7.5 - Optimizing the Deposition Rate for a Silicon Wafer Process
Buckner, Cammenga, and Weber (1993) ran a two factor, spherical ccd to optimize the deposition rate fro a tungsten film on a silicon wafer in terms of the process temperature and the ratio of H_2 to W F_6 in the reaction atmosphere. The results

using the design variables, follow.

x_1	x_2	y
-1	-1	3663
1	-1	9393
-1	1	5602
1	1	12488
-1.414	0	1984
1.414	0	12603
0	-1.414	5007
0	1.414	10310
0	0	8979
0	0	8960
0	0	8979

Follow these four steps to construct this central composite design.

Step 1. Obtain a part of the design matrix.

Choose **Stat>DOE>Create RS Design.** Darken the Central Composite Type of Design option button. Place 2 in the Number of factors: text box. Select the design.

Select Designs. Highlight Full with 13 Runs. Choose **OK.**

Obtain a standard order for the design.

Choose Options. Remove the check from the Randomize runs checkbox. Choose **OK.**

Choose **OK.**

Step 2. Delete some rows of the design matrix.

Observe that we have only eleven experimental results corresponding to the first eleven elements of the design matrix. Highlight rows 12 and 13, and delete those rows.

Step 3. Enter data.

Enter the results of the experiment ($y_i's$) in column C6. Name column C6 as Y.

Step 4. Estimate the effects.

Choose **Stat>DOE>Analyze RS Design.** Place Y in the Responses: text box. Choose **OK.**

The Minitab Output

Response Surface Regression

The analysis was done using coded units.

Estimated Regression Coefficients for Y

Term	Coef	StDev	T	P
Constant	8972.7	334.1	26.860	0.000
A	3454.2	204.6	16.886	0.000
B	1566.7	204.6	7.659	0.001
A*A	-762.0	243.5	-3.129	0.026
B*B	-579.5	243.5	-2.380	0.063
A*B	289.0	289.3	0.999	0.364

S = 578.6   R-Sq = 98.6%   R-Sq(adj) = 97.2%

Analysis of Variance for Y

Source	DF	Seq SS	Adj SS	Adj MS	F	P
Regression	5	119481105	119481105	23896221	71.38	0.000
Linear	2	115087835	115087835	57543918	171.89	0.000
Square	2	4059186	4059186	2029593	6.06	0.046
Interaction	1	334084	334084	334084	1.00	0.364
Residual Error	5	1673864	1673864	334773		
Lack-of-Fit	3	1673623	1673623	557874	5E+03	0.000
Pure Error	2	241	241	120		
Total	10	121154969				

Figure 7.15

The analysis, as shown in the Minitab output in Figure 7.15, suggests the following prediction equation

$$\hat{y} = 8972 + 3454x_1 + 1567x_2 - 762x_1^2 - 579x_2^2.$$

We can obtain a surface plot of this model using Minitab.

Follow this step to obtain the surface plot.

Choose **Stat>DOE>RS Plots.** Place a check in the Surface (wireframe) plot checkbox.

Setup the display.

Choose Setup. Darken the Display plots using: Coded units option button. Choose **OK.**

Choose **OK.**

The Minitab Output

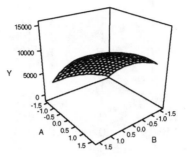

Figure 7.16

The Minitab output, as shown in Figure 7.16, provides a three-dimensional response surface to determine optimum conditions. The plot suggests that we should increase both the pressure and the ration of H_2 to WF_6 to maximize the deposition rate.

Exercises

7.12 Consider Example 7.7 in the text, the springs with cracks experiment.
   a. Construct a face centered cube ccd for this situation in both the design and the natural variables.
   b. Construct a spherical ccd for this situation in both the design and the natural variables.

7.13 Consider Example 7.10, the oil extraction from peanuts experiment.
   a. Construct a face centered cube ccd using all five factors in both the design and the natural variables.
   b. Construct a rotatable ccd using all five factors in both the design and the natural variables.

7.14 (EXO731)Derringer and Suich (1980) used a ccd to optimize an abrasion index for a tire tread compound in terms of three factors: x_1, hydrated silica level; x_2, silane coupling agent level; and x_3, sulfur level. The actual results follow.

Perform a thorough analysis of the results including residual plots.

x_1	x_2	x_3	y
-1	-1	1	102
1	-1	-1	120
-1	1	-1	117
1	1	1	198
-1	-1	-1	103
1	-1	1	132
-1	1	1	132
1	1	-1	139
-1.633	0	0	102
1.633	0	0	154
0	-1.633	0	96
0	1.633	0	163
0	0	-1.633	116
0	0	1.633	153
0	0	0	133
0	0	0	133
0	0	0	140
0	0	0	142
0	0	0	145
0	0	0	142

Follow these steps to begin your analysis of this central composite design.

Step 1. Obtain the design matrix and the data.
Choose **File**>**Open worksheet.** Retrieve the file EXO731.MTP.

Step 2. Estimate the effects.
Choose **Stat**>**DOE**>**Analyze RS Design.** Place Y in the
Responses: text box. Choose **OK.**